T0281377

PySpark Recipes

A Problem-Solution Approach
with PySpark2

Raju Kumar Mishra

Apress®

PySpark Recipes

Raju Kumar Mishra
Bangalore, Karnataka, India

ISBN-13 (pbk): 978-1-4842-3140-1   ISBN-13 (electronic): 978-1-4842-3141-8
https://doi.org/10.1007/978-1-4842-3141-8

Library of Congress Control Number: 2017962438

Copyright © 2018 by Raju Kumar Mishra

This work is subject to copyright. All rights are reserved by the Publisher, whether the whole or part of the material is concerned, specifically the rights of translation, reprinting, reuse of illustrations, recitation, broadcasting, reproduction on microfilms or in any other physical way, and transmission or information storage and retrieval, electronic adaptation, computer software, or by similar or dissimilar methodology now known or hereafter developed.

Trademarked names, logos, and images may appear in this book. Rather than use a trademark symbol with every occurrence of a trademarked name, logo, or image, we use the names, logos, and images only in an editorial fashion and to the benefit of the trademark owner, with no intention of infringement of the trademark.

The use in this publication of trade names, trademarks, service marks, and similar terms, even if they are not identified as such, is not to be taken as an expression of opinion as to whether or not they are subject to proprietary rights.

While the advice and information in this book are believed to be true and accurate at the date of publication, neither the authors nor the editors nor the publisher can accept any legal responsibility for any errors or omissions that may be made. The publisher makes no warranty, express or implied, with respect to the material contained herein.

Cover image by Freepik (`www.freepik.com`)

Managing Director: Welmoed Spahr
Editorial Director: Todd Green
Acquisitions Editor: Celestin Suresh John
Development Editor: Laura Berendson
Technical Reviewer: Sundar Rajan
Coordinating Editor: Sanchita Mandal
Copy Editor: Sharon Wilkey
Compositor: SPi Global
Indexer: SPi Global
Artist: SPi Global

Distributed to the book trade worldwide by Springer Science + Business Media New York, 233 Spring Street, 6th Floor, New York, NY 10013. Phone 1-800-SPRINGER, fax (201) 348-4505, e-mail orders-ny@springer-sbm.com, or visit `www.springeronline.com`. Apress Media, LLC is a California LLC, and the sole member (owner) is Springer Science + Business Media Finance Inc (SSBM Finance Inc). SSBM Finance Inc is a **Delaware** corporation.

For information on translations, please e-mail rights@apress.com, or visit `www.apress.com/rights-permissions`.

Apress titles may be purchased in bulk for academic, corporate, or promotional use. eBook versions and licenses are also available for most titles. For more information, reference our Print and eBook Bulk Sales web page at `www.apress.com/bulk-sales`.

Any source code or other supplementary material referenced by the author in this book is available to readers on GitHub via the book's product page, located at `www.apress.com`. For more detailed information, please visit `www.apress.com/source-code`.

Printed on acid-free paper

To the Almighty, who guides me in every aspect of my life.
And to my mother, Smt. Savitri Mishra, and
my lovely wife, Smt. Smita Rani Pathak.

Contents

About the Author

Raju Kumar Mishra has a strong interest in data science and systems that have the capability of handling large amounts of data and operating complex mathematical models through computational programming. He was inspired to pursue a Master of Technology degree in computational sciences from the Indian Institute of Science in Bangalore, India. Raju primarily works in the areas of data science and its various applications. Working as a corporate trainer, he has developed unique insights that help him in teaching and explaining complex ideas with ease. Raju is also a data science consultant who solves complex industrial problems. He works on programming tools such as R, Python, scikit-learn, Statsmodels, Hadoop, Hive, Pig, Spark, and many others.

About the Technical Reviewer

Sundar Rajan Raman is an artificial intelligence practitioner currently working for Bank of America. He holds a Bachelor of Technology degree from the National Institute of Technology in India. Being a seasoned Java and J2EE programmer, he has worked at companies such as AT&T, Singtel, and Deutsche Bank. He is a messaging platform specialist with vast experience on SonicMQ, WebSphere MQ, and TIBCO software, with respective certifications. His current focus is on artificial intelligence, including machine learning and neural networks. More information is available at https://in.linkedin.com/pub/sundar-rajan-raman/7/905/488.

   I would like to thank my wife, Hema, and my daughter, Shriya, for their patience during the review process.

Acknowledgments

My heartiest thanks to the Almighty. I also would like to thank my mother, Smt. Savitri Mishra; my sisters, Mitan and Priya; my cousins, Suchitra and Chandni; and my maternal uncle, Shyam Bihari Pandey; for their support and encouragement. I am very grateful to my sweet and beautiful wife, Smt. Smita Rani Pathak, for her continuous encouragement and love while I was writing this book. I thank my brother-in-law, Mr. Prafull Chandra Pandey, for his encouragement to write this book. I am very thankful to my sisters-in-law, Rinky, Reena, Kshama, Charu, Dhriti, Kriti, and Jyoti for their encouragement as well. I am grateful to Anurag Pal Sehgal, Saurabh Gupta, Devendra Mani Tripathi, and all my friends. Last but not least, thanks to Coordinating Editor Sanchita Mandal, Acquisitions Editor Celestin Suresh John, and Development Editor Laura Berendson at Apress; without them, this book would not have been possible.

Introduction

This book will take you on an interesting journey to learn about PySpark and big data through a problem-solution approach. Every problem is followed by a detailed, step-by-step answer, which will improve your thought process for solving big data problems with PySpark. This book is divided into nine chapters. Here's a brief description of each chapter:

Chapter 1, "The Era of Big Data, Hadoop, and Other Big Data Processing Frameworks," covers many big data processing tools such as Apache Hadoop, Apache Pig, Apache Hive, and Apache Spark. The shortcomings of Hadoop and the evolution of Spark are discussed. Apache Kafka is explained as a publish-subscribe system. This chapter also sheds light on HBase, a NoSQL database.

Chapter 2, "Installation," will take you to the real battleground. You'll learn how to install many big data processing tools such as Hadoop, Hive, Spark, Apache Mesos, and Apache HBase.

Chapter 3, "Introduction to Python and NumPy," is for newcomers to Python. You will learn about the basics of Python and NumPy by following a problem-solution approach. Problems in this chapter are data-science oriented.

Chapter 4, "Spark Architecture and the Resilient Distributed Dataset," explains the architecture of Spark and introduces resilient distributed datasets. You'll learn about creating RDDs and using data-analysis algorithms for data aggregation, data filtering, and set operations on RDDs.

Chapter 5, "The Power of Pairs: Paired RDD," shows how to create paired RDDs and how to perform data aggregation, data joining, and other algorithms on these paired RDDs.

Chapter 6, "I/O in PySpark," will teach you how to read data from various types of files and save the result as an RDD.

Chapter 7, "Optimizing PySpark and PySpark Streaming," is one of the most important chapters. You will start by optimizing a page-rank algorithm. Then you'll implement a k-nearest neighbors algorithm and optimize it by using broadcast variables provided by the PySpark framework. Learning PySpark Streaming will finally lead us into integrating Apache Kafka with the PySpark Streaming framework.

Chapter 8, "PySparkSQL," is paradise for readers who use SQL. But newcomers will also learn PySparkSQL in order to write SQL-like queries on DataFrames by using a problem-solution approach. Apart from DataFrames, we will also implement the graph algorithms breadth-first search and page rank by using the GraphFrames library.

Chapter 9, "PySpark MLlib and Linear Regression," describes PySpark's machine-learning library, MLlib. You will see many recipes on various data structures provided by PySpark MLlib. You'll also implement linear regression. Recipes on lasso and ridge regression are included in the chapter.

CHAPTER 1

■ ■ ■

The Era of Big Data, Hadoop, and Other Big Data Processing Frameworks

When I first joined Orkut, I was happy. With Orkut, I had a new platform enabling me get to know the people around me, including their thoughts, their views, their purchases, and the places they visited. We were all gaining more knowledge than ever before and felt more connected to the people around us. Uploading pictures helped us share good ideas of places to visit. I was becoming more and more addicted to understanding and expressing sentiments. After a few years, I joined Facebook. And day by day, I was introduced to what became an infinite amount of information from all over world. Next, I started purchasing items online, and I liked it more than shopping offline. I could easily get a lot of information about products, and I could compare prices and features. And I wasn't the only one; millions of people were feeling the same way about the Web.

More and more data was flooding in from every corner of the world to the Web. And thanks to all those inventions related to data storage systems, people could store this huge inflow of data.

More and more users joined the Web from all over the world, and therefore increased the amount of data being added to these storage systems. This data was in the form of opinions, pictures, videos, and other forms of data too. This data deluge forced users to adopt distributed systems. Distributed systems require distributed programming. And we also know that distributed systems require extra care for fault-tolerance and efficient algorithms. Distributed systems always need two things: reliability of the system and availability of all its components.

Apache Hadoop was introduced, ensuring efficient computation and fault-tolerance for distributed systems. Mainly, it concentrated on reliability and availability. Because Apache Hadoop was easy to program, many people became interested in big data. Big data became a popular topic for discussion everywhere. E-commerce companies wanted to know more about their customers, and the health-care industry was interested in gaining insights from the data collected, for example. More data metrics were defined. More data points started to be collected.

© Raju Kumar Mishra 2018
R. K. Mishra, *PySpark Recipes*, https://doi.org/10.1007/978-1-4842-3141-8_1

Many open source big data tools emerged, including Apache Tez and Apache Storm. This was also a time that many NoSQL databases emerged to deal with this huge data inflow. Apache Spark also evolved as a distributed system and became very popular during this time.

In this chapter, we are going to discuss big data as well as Hadoop as a distributed system for processing big data. In covering the components of Hadoop, we will also discuss Hadoop ecosystem frameworks such as Apache Hive and Apache Pig. The usefulness of the components of the Hadoop ecosystem is also discussed to give you an overview. Throwing light on some of the shortcomings of Hadoop will give you background on the development of Apache Spark. The chapter will then move through a description of Apache Spark. We will also discuss various cluster managers that work with Apache Spark. The chapter wouldn't be complete without discussing NoSQL, so discussion on the NoSQL database HBase is also included. Sometimes we read data from a relational database management system (RDBMS); this chapter discusses PostgreSQL.

Big Data

Big data is one of the hot topics of this era. But what is big data? Big data describes a dataset that is huge and increasing with amazing speed. Apart from this volume and velocity, big data is also characterized by its variety of data and veracity. Let's explore these terms—volume, velocity, variety, and veracity—in detail. These are also known as the *4V characteristics* of big data, as illustrated in Figure 1-1.

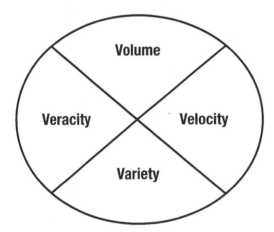

Figure 1-1. *Characteristcis of big data*

Volume

The *volume* specifies the amount of data to be processed. A large amount of data requires large machines or distributed systems. And the time required for computation will also increase with the volume of data. So it's better to go for a distributed system, if we can parallelize our computation. Volume might be of structured data, unstructured data, or any data. If we have unstructured data, the situation becomes more complex and computing intensive. You might wonder, how big is big? What volume of data should be classified as big data? This is again a debatable question. But in general, we can say that an amount of data that we can't handle via a conventional system can be considered big data.

Velocity

Every organization is becoming more and more data conscious. A lot of data is collected every moment. This means that the *velocity* of data—the speed of the data flow and of data processing—is also increasing. How will a single system be able to handle this velocity? The problem becomes complex when we have to analyze a large inflow of data in real time. Each day, systems are being developed to deal with this huge inflow of data.

Variety

Sometimes the *variety* of data adds enough complexity that conventional data analysis systems can't analyze data well. What do we mean by *variety*? You might think data is just data. But this is not the case. Image data is different from simple tabular data, for example, because of the way it is organized and saved. In addition, an infinite number of file systems are available, and every file system requires a different way of dealing with it. Reading and writing a JSON file, for instance, will be different from the way we deal with a CSV file. Nowadays, a data scientist has to handle a combination of these data types. The data you are going to deal with might be a combination of pictures, videos, and text. This variety of data makes big data more complex to analyze.

Veracity

Can you imagine a logically incorrect computer program resulting in the correct output? Of course not. Similarly, data that is not accurate is going to provide misleading results. The *veracity* of data is one of the important concerns related to big data. When we consider the condition of big data, we have to think about any abnormalities in the data.

Hadoop

Hadoop is a distributed and scalable framework for solving big data problems. Hadoop, developed by Doug Cutting and Mark Cafarella, is written in Java. It can be installed on a cluster of commodity hardware, and it scales horizontally on distributed systems. Easy to program Inspiration from Google research paper Hadoop was developed. Hadoop's capability to work on commodity hardware makes it cost-effective. If we are working on commodity hardware, fault-tolerance is an inevitable issue. But Hadoop provides a fault-tolerant system for data storage and computation, and this fault-tolerant capability has made Hadoop popular.

Hadoop has two components, as illustrated in Figure 1-2. The first component is the Hadoop Distributed File System (HDFS). The second component is MapReduce. HDFS is for distributed data storage, and MapReduce is for performing computation on the data stored in HDFS.

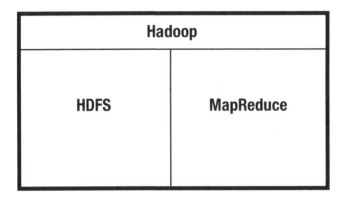

Figure 1-2. *Hadoop components*

HDFS

HDFS is used to store large amounts of data in a distributed and fault-tolerant fashion. HDFS is written in Java and runs on commodity hardware. It was inspired by a Google research paper about the Google File System (GFS). It is a write-once and read-many-times system that's effective for large amounts of data.

HDFS comprises two components: NameNode and DataNode. These two components are Java daemon processes. A NameNode, which maintains metadata of files distributed on a cluster, works as the master for many DataNodes. HDFS divides a large file into small blocks and saves the blocks on different DataNodes. The actual file data blocks reside on DataNodes.

HDFS provides a set of Unix shell-like commands to deal with it. But we can use the Java file system API provided by HDFS to work at a finer level on large files. Fault-tolerance is implemented by using replications of data blocks.

We can access the HDFS files by using a single-thread process and also in parallel. HDFS provides a useful utility, distcp, which is generally used to transfer data in parallel from one HDFS system to another. It copies data by using parallel map tasks. You can see the HDFS components in Figure 1-3.

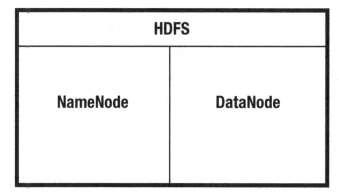

Figure 1-3. *Components of HDFS*

MapReduce

The Map-Reduce model of computation first appeared in a Google research paper. This research paper was implemented in Hadoop as Hadoop's MapReduce. Hadoop's MapReduce is the computation engine of the Hadoop framework, which performs computations on the distributed data in HDFS. MapReduce is horizontally scalable on distributed systems of commodity hardware. It also scales for large problems. In MapReduce, the solution is broken into two phases: the map phase and the reduce phase. In the map phase, a chunk of data is processed, and in the reduce phase, an aggregation or a reduction operation is run on the result of the map phase. Hadoop's MapReduce framework is written in Java.

MapReduce uses a master/slave model. In Hadoop 1, this map-reduce computation was managed by two daemon processes: Jobtracker and Tasktracker. Jobtracker is a master process that deals with many Tasktrackers. There's no need to say that Tasktracker is a slave to Jobtracker. But in Hadoop 2, Jobtracker and Tasktracker were replaced by YARN.

Because we know that Hadoop's MapReduce framework is written in Java, we can write our MapReduce code by using an API provided by the framework and programmed in Java. The Hadoop streaming module gives further power so that a person knowing another programming language (such as Python or Ruby) can program MapReduce.

MapReduce algorithms are good for many algorithms. Many machine-learning algorithms are implemented as Apache Mahout. Mahout used to run on Hadoop as Pig and Hive.

But MapReduce wasn't very good for iterative algorithms. At the end of every Hadoop job, MapReduce will save the data to HDFS and read it back again for the next job. We know that reading and writing data to a file is one of the costliest activities. Apache Spark mitigated this shortcoming of MapReduce by providing in-memory data persisting and computation.

▪ **Note**   You can read more about MapReduce and Mahout at the following web pages:

www.usenix.org/legacy/publications/library/proceedings/osdi04/tech/
full_papers/dean/dean_html/index.html

https://mahout.apache.org/users/basics/quickstart.html

Apache Hive

The world of computer science is one of abstraction. Everyone knows that all data ultimately exists in the form of bits. Programming languages such as C enable us to avoid programming in machine-level language. The C language provides an abstraction over machine and assembly language. More abstraction is provided by other high-level languages. Structured Query Language (SQL) is one of the abstractions. SQL is widely used all over the world by many data modeling experts. Hadoop is good for analysis of big data. So how can a large population knowing SQL utilize the power of Hadoop computational power on big data? In order to write Hadoop's MapReduce program, users must know a programming language that can be used to program Hadoop's MapReduce.

In the real world, day-to-day problems follow patterns. In data analysis, some problems are common, such as manipulating data, handling missing values, transforming data, and summarizing data. Writing MapReduce code for these day-to-day problems is head-spinning work for a nonprogrammer. Writing code to solve a problem is not a very intelligent thing. But writing efficient code that has performance scalability and can be extended is something that is valuable. Having this problem in mind, *Apache Hive* was developed at Facebook, so that general problems can be solved without writing MapReduce code.

According to the Hive wiki, "Hive is a data warehousing infrastructure based on Apache Hadoop." Hive has its own SQL dialect, which is known as *Hive Query Language* (abbreviated as HiveQL or HQL). Using HiveQL, Hive can query data in HDFS. Hive can run not only on HDFS, but also on Spark and other big data frameworks such as Apache Tez.

Hive provides the user an abstraction that is like a relational database management system for structured data in HDFS. We can create tables and run SQL-like queries on them. Hive saves the table schema in an RDBMS. Apache Derby is the default RDBMS, which is shipped with the Apache Hive distribution. Apache Derby has been fully written in Java; this open source RDBMS comes with the Apache License, Version 2.0.

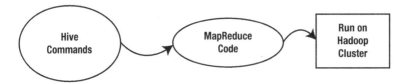

Figure 1-4. *Code execution flow in Apache Hive*

HiveQL commands are transformed into Hadoop's MapReduce code, and then it runs on Hadoop cluster. You can see the Hive command execution flow in Figure 1-4.

A person knowing SQL can easily learn Apache Hive and HiveQL and can use the benefits of storage and the computation power of Hadoop in their day-to-day data analysis of big data. HiveQL is also supported by PySparkSQL. We can run HiveQL commands in PySparkSQL. Apart from executing HiveQL queries, we can also read data from Hive directly to PySparkSQL and write results to Hive.

■ **Note**   You can read more about Hive and the Apache Derby RDBMS at the following web pages:

https://cwiki.apache.org/confluence/display/Hive/Tutorial

https://db.apache.org/derby/

Apache Pig

Apache Pig is data-flow framework for performing data-analysis algorithms on huge amounts of data. It was developed by Yahoo!, open sourced to the Apache Software Foundation, and is now available under the Apache License, Version 2.0. The pig programming language is a Pig Latin scripting language. Pig is loosely connected to Hadoop, which means that we can connect it to Hadoop and perform analysis. But Pig can be used with other tools such as Apache Tez and Apache Spark.

Apache Hive is used as reporting tool, whereas Apache Pig is used as an extract, transform, and load (ETL) tool. We can extend the functionality of Pig by using user-defined functions (UDFs). User-defined functions can be written in many languages, including Java, Python, Ruby, JavaScript, Groovy, and Jython.

Apache Pig uses HDFS to read and store the data, and Hadoop's MapReduce to execute the data-science algorithms. Apache Pig is similar to Apache Hive in using the Hadoop cluster. As Figure 1-5 depicts, on Hadoop, Pig Latin commands are first transformed into Hadoop's MapReduce code. And then the transformed MapReduce code runs on the Hadoop cluster.

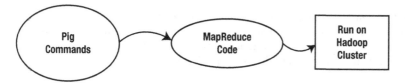

Figure 1-5. Code execution flow in Apache Pig

The best part of Pig is that the code is optimized and tested to work for day-to-day problems. A user can directly install Pig and start using it. Pig provides a Grunt shell to run interactive Pig commands, so anyone who knows Pig Latin can enjoy the benefits of HDFS and MapReduce, without knowing an advanced programming language such as Java or Python.

■ **Note**   You can read more about Apache Pig at the following sites:

http://pig.apache.org/docs/

https://en.wikipedia.org/wiki/Pig_(programming_tool)

https://cwiki.apache.org/confluence/display/PIG/Index

Apache Kafka

Apache Kafka is a publish-subscribe, distributed messaging platform. It was developed at LinkedIn and later open sourced to the Apache Foundation. It is fault-tolerant, scalable, and fast. A *message*, in Kafka terms, is the smallest unit of data that can flow from a producer to a consumer through a Kafka server, and that can be persisted and used at a later time. You might be confused about the terms *producer* and *consumer*. We are going to discuss these terms soon. Another key term we are going to use in the context of Kafka is *topic*. A topic is stream of messages of a similar category. Kafka comes with a built-in API, which developers can use to build their applications. We are the ones who define the topic. Now let's discuss the three main components of Apache Kafka.

Producer

A Kafka *producer* produces the message to a Kafka topic. It can publish data to more than one topic.

Broker

The *broker* is the main Kafka server that runs on a dedicated machine. Messages are pushed to the broker by the producer. The broker persists topics in different partitions, and these partitions are replicated to different brokers to deal with faults. The broker is stateless, so the consumer has to track the message it has consumed.

Consumer

A *consumer* fetches messages from the Kafka broker. Remember, it fetches the messages; the Kafka broker doesn't push messages to the consumer; rather, the consumer pulls data from the Kafka broker. Consumers are subscribed to one or more topics on the Kafka broker, and they read the messages. The consumer also keeps tracks of all the messages that it has already consumed. Data is persisted in a broker for a specified time. If the consumer fails, it can fetch the data after its restart.

Figure 1-6 explains the message flow of Apache Kafka. The producer publishes a message to the topic. Then the consumer pulls data from the broker. In between publishing and pulling, the message is persisted by the Kafka broker.

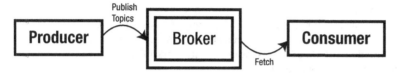

Figure 1-6. *Apache Kafka message flow*

We will integrate Apache Kafka with PySpark in Chapter 7, which discusses Kafka further.

■ **Note**   You can read more about Apache Kafka at the following sites:

https://kafka.apache.org/documentation/

https://kafka.apache.org/quickstart

Apache Spark

Apache Spark is a general-purpose, distributed programming framework. It is considered very good for iterative as well as batch processing of data. Developed at the AMPLab at the University of California, Berkeley, Spark is now open source software that provides an in-memory computation framework. On the one hand, it is good for batch processing; on the other hand, it works well with real-time (or, better to say, near-real-time) data. Machine learning and graph algorithms are iterative. Where Spark do magic. According to its research paper, it is approximately 100 times faster than its peer, Hadoop. Data can be cached in memory. Caching intermediate data in iterative algorithms provides amazingly fast processing speed. Spark can be programmed with Java, Scala, Python, and R.

If anyone is considering Spark as an improved Hadoop, then to some extent, that is fine in my view. Because we can implement a MapReduce algorithm in Spark, Spark uses the benefit of HDFS; this means Spark can read data from HDFS and store data to HDFS too, and Spark handles iterative computation efficiently because data can be persisted in memory. Apart from in-memory computation, Spark is good for interactive data analysis.

We are going to study Apache Spark with Python. This is also known as *PySpark*. PySpark comes with many libraries for writing efficient programs, and there are some external libraries as well. Here are some of them:

- *PySparkSQL*: A PySpark library to apply SQL-like analysis on a huge amount of structured or semistructured data. We can also use SQL queries with PySparkSQL. We can connect it to Apache Hive, and HiveQL can be applied too. PySparkSQL is a wrapper over the PySpark core. PySparkSQL introduced the DataFrame, which is a tabular representation of structured data that is like a table in a relational database management system. Another data abstraction, the DataSet, was introduced in Spark 1.6, but it does not work with PySparkSQL.

- *MLlib*: MLlib is a wrapper over the PySpark core that deals with machine-learning algorithms. The machine-learning API provided by the MLlib library is easy to use. MLlib supports many machine-learning algorithms for classification, clustering, text analysis, and more.

- *GraphFrames*: The GraphFrames library provides a set of APIs for performing graph analysis efficiently, using the PySpark core and PySparkSQL. At the time of this writing, DataFrames is an external library. You have to download and install it separately. We are going to perform graph analysis in Chapter 8.

Cluster Managers

In a distributed system, a job or application is broken into different tasks, which can run in parallel on different machines of the cluster. A task, while running, needs resources such as memory and a processor. The most important part is that if a machine fails, you then have to reschedule the task on another machine. The distributed system generally faces scalability problems due to mismanagement of resources. As another scenario, say a job is already running on a cluster. Another person wants to run another job. The second job has to wait until the first is finished. But in this way, we are not utilizing the resources optimally. This resource management is easy to explain but difficult to implement on a distributed system.

Cluster managers were developed to manage cluster resources optimally. There are three cluster managers available for Spark: Standalone, Apache Mesos, and YARN. The best part of these cluster managers is that they provide an abstraction layer between the user and the cluster. The user feels like he's working on a single machine, while in reality he's working on a cluster, due to the abstraction provided by cluster managers. Cluster managers schedule cluster resources to running applications.

Standalone Cluster Manager

Apache Spark is shipped with the Standalone Cluster Manager. It provides a master/slave architecture to the Spark cluster. It is Spark's only cluster manager. You can run only Spark applications when using the Standalone Cluster Manager. Its components are the master and workers. Workers are the slaves to the master process. Standalone is the simplest cluster manager. Spark Standalone Cluster Manager can be configured using scripts in the sbin directory of Spark. We will configure Spark Standalone Cluster Manager in the coming chapters and will deploy PySpark applications by using Standalone Cluster Manager.

Apache Mesos Cluster Manager

Apache Mesos is a general-purpose cluster manager. It was developed at the University of California, Berkeley, AMPLab. Apache Mesos helps distributed solutions scale efficiently. You can run different applications using different frameworks on the same cluster when using Mesos. What do I mean by *different applications using different frameworks*? I mean that we can run a Hadoop application and a Spark application simultaneously on Mesos. While multiple applications are running on Mesos, they share the resources of the cluster. The two important components of Apache Mesos are master and slaves. It has a master/slave architecture similar to Spark Standalone Cluster Manager. The applications running on Mesos are known as the framework. Slaves inform the master about the resources available to it as a resource offer. Slave machines provides resource offers periodically. The allocation module of the master server decides the framework that will get the resources.

YARN Cluster Manager

YARN stands for *Yet Another Resource Negotiator*. YARN was introduced in Hadoop 2 to scale Hadoop; resource management and job management were separated. Separating these two components made Hadoop scale better. YARN's main components are ResourceManager, ApplicationMaster, and NodeManager. There is one global ResourceManager, and many NodeManagers will be running per cluster. NodeManagers are slaves to the ResourceManager. The Scheduler, which is a component of ResourceManager, allocates resources for different applications working on the cluster. The best part is, we can run a Spark application and any other applications such as Hadoop or MPI simultaneously on clusters managed by YARN. There is one ApplicationMaster per application, which deals with the task running in parallel on a distributed system. Remember, Hadoop and Spark have their own kinds of ApplicationMaster.

■ **Note**   You can read more about Standalone, Apache Mesos, and YARN cluster managers at the following web pages:

https://spark.apache.org/docs/2.0.0/spark-standalone.html

https://spark.apache.org/docs/2.0.0/running-on-mesos.html

https://spark.apache.org/docs/2.0.0/running-on-yarn.html

PostgreSQL

Relational database management systems are till very frequent in different organizations. What is the meaning or *relational* here? It means *tables*. PostgreSQL is an RDBMS. It runs on nearly all major operating systems, including Microsoft Windows, Unix-based operating systems, macOS, and many more. It is open source software, and the code is available under the PostgreSQL license. Therefore, you can use it freely and modify it according to your requirements.

PostgreSQL databases can be connected through other programming languages such as Java, Perl, Python, C, and C++ and through various programming interfaces. It can be also be programmed using a procedural programming language, Procedural Language/PostgreSQL (PL/pgSQL), which is similar to PL/SQL. The user can add custom functions to this database. We can write our custom functions in C/C++ and other programming languages. We can read data from PostgreSQL from PySparkSQL by using Java Database Connectivity (JDBC) connectors. In upcoming chapters, we are going to read data tables from PostgreSQL by using PySparkSQL. We are also going to explore more facets of PostgreSQL in upcoming chapters.

PostgreSQL follows the ACID (Atomicity, Consistency, Isolation, and Durability) principles. It comes with many features, and some might be unique to PostgreSQL itself. It supports updatable views, transactional integrity, complex queries, triggers, and other features. PostgreSQL performs its concurrency management by using a multiversion concurrency control model.

There is a large community of support if you find a problem while using PostgreSQL. PostgreSQL has been designed and developed to be extensible.

■ **Note**   If you want to learn PostgreSQL in depth, the following links will be helpful to you:

https://wiki.postgresql.org/wiki/Main_Page

https://en.wikipedia.org/wiki/PostgreSQL

https://en.wikipedia.org/wiki/Multiversion_concurrency_control

http://postgresguide.com/

HBase

HBase is an open source, distributed, NoSQL database. When I say *NoSQL*, you might consider it schemaless. And you're right, to a certain extent, but not completely. At the time that you define a table, you have to mention the column family, so the database is not fully schemaless. We are going to create an HBase table in this section so you can understand this semi-schemaless property. HBase is a column-oriented database. You might wonder what that means. Let me explain: in column-oriented databases, data is saved columnwise.

We are going to install HBase in the next chapter, but for now, let me show how a table is created and how data is put inside the tables. You can apply all these commands after installing HBase on your system. In the coming chapter, we are going to read the same data table by using PySpark.

```
hbase(main):001:0> list
TABLE
0 row(s) in 0.1750 seconds

=> []
hbase(main):002:0> create 'pysparkBookTable','btcf1','btcf2'
0 row(s) in 2.2750 seconds

=> Hbase::Table - pysparkBookTable
hbase(main):003:0> list
TABLE
pysparkBookTable
1 row(s) in 0.0190 seconds

=> ["pysparkBookTable"]
hbase(main):004:0> put 'pysparkBookTable', '00001', 'btcf1:btc1','c11'
0 row(s) in 0.1680 seconds

hbase(main):005:0> put 'pysparkBookTable', '00001', 'btcf2:btc2','c21'
0 row(s) in 0.0240 seconds
hbase(main):006:0> put 'pysparkBookTable', '00002', 'btcf1:btc1','c12'
0 row(s) in 0.0150 seconds

hbase(main):007:0> put 'pysparkBookTable', '00002', 'btcf2:btc2','c22'
0 row(s) in 0.0070 seconds

hbase(main):008:0> put 'pysparkBookTable', '00003', 'btcf1:btc1','c13'
0 row(s) in 0.0080 seconds

hbase(main):009:0> put 'pysparkBookTable', '00003', 'btcf2:btc2','c23'
0 row(s) in 0.0060 seconds

hbase(main):010:0>  put 'pysparkBookTable', '00004', 'btcf1:btc1','c14'
0 row(s) in 0.0240 seconds

hbase(main):011:0>  put 'pysparkBookTable', '00004', 'btcf2:btc2','c24'
0 row(s) in 0.0280 seconds
```

```
hbase(main):012:0> scan 'pysparkBookTable'
ROW                      COLUMN+CELL
 00001                   column=btcf1:btc1, timestamp=1496715394968, value=c11
 00001                   column=btcf2:btc2, timestamp=1496715408865, value=c21
 00002                   column=btcf1:btc1, timestamp=1496715423206, value=c12
 00002                   column=btcf2:btc2, timestamp=1496715436087, value=c22
 00003                   column=btcf1:btc1, timestamp=1496715450562, value=c13
 00003                   column=btcf2:btc2, timestamp=1496715463134, value=c23
 00004                   column=btcf1:btc1, timestamp=1496715503014, value=c14
 00004                   column=btcf2:btc2, timestamp=1496715516864, value=c24
4 row(s) in 0.0770 seconds
```

■ **Note**   You can get a lot of information about HBase at `https://hbase.apache.org/`.

Spark can be used with three cluster managers: Standalone, Apache Mesos, and YARN. Standalone cluster manager is shipped with Spark and it is Spark only cluster manager. With Apache Mesos and YARN, we can run heterogeneous applications.

CHAPTER 2

■ ■ ■

Installation

In the upcoming chapters, we are going to solve many problems by using PySpark. PySpark also interacts with many other big data frameworks to provide end-to-end solutions. PySpark might read data from HDFS, NoSQL databases, or a relational database management system (RDBMS). After data analysis, we can also save the results into HDFS or databases.

This chapter covers all the software installations that are required to go through this book. We are going to install all the required big data frameworks on the CentOS operating system. CentOS is an enterprise-class operating system. It is free to use and easily available. You can download CentOS from www.centos.org/download/ and then install it on a virtual machine.

This chapter covers the following recipes:

- Recipe 2-1. Install Hadoop on a single machine

- Recipe 2-2. Install Spark on a single machine

- Recipe 2-3. Use the PySpark shell

- Recipe 2-4. Install Hive on a single machine

- Recipe 2-5. Install PostgreSQL

- Recipe 2-6. Configure the Hive metastore on PostgreSQL

- Recipe 2-7. Connect PySpark to Hive

- Recipe 2-8. Install Apache Mesos

- Recipe 2-9. Install HBase

I suggest that you install every piece of software on your own. It is a good exercise and will give you a deeper understanding of the components of each software package.

© Raju Kumar Mishra 2018
R. K. Mishra, *PySpark Recipes*, https://doi.org/10.1007/978-1-4842-3141-8_2

Recipe 2-1. Install Hadoop on a Single Machine

Problem

You want to install Hadoop on a single machine.

Solution

You might be thinking, Why are we installing Hadoop while we are learning PySpark? Are we going to use Hadoop MapReduce as a distributed framework for our problem solving? The answer is, Not at all. We are going to use two components of Hadoop: HDFS and YARN—HDFS for data storage and YARN as cluster manager. Installation of Hadoop requires you to download it and configure it.

How It Works

Follow these steps to complete the installation.

Step 2-1-1. Creating a New CentOS User

In this step, we'll create a new user. You might be thinking, Why a new user? Why can't we install Hadoop on an existing user? The reason is that we want to provide a dedicated user for all the big data frameworks. With the following lines of code, we create the user pysparkbook:

```
[root@localhost pyspark]# adduser pysparkbook
[root@localhost pyspark]# passwd pysparkbook
```

The output is as follows:

```
Changing password for user pysparkbook.
New password:
passwd: all authentication tokens updated successfully.
```

In the preceding code, you can see that the command adduser has been used to create or add a user. The Linux command passwd has been used to provide a password to our new user pysparkbook.

After creating the user, we have to add it to sudo. Sudo stands for *superuser do*. Using sudo, we can run any code as a super user. Sudo is used to install software.

Step 2-1-2. Creating a new CentOS user

A new user is created. You might be thinking why new user. Why cant we install Hadoop in existing user. The reason behind that is, we want to provide a dedicated user for all the big data frameworks. In following lines of code we are going to create a user "pysparkbook".

```
[pyspark@localhost ~]$ su root
[root@localhost pyspark]# adduser pysparkbook
[root@localhost pyspark]# passwd pysparkbook
```

Output:

```
Changing password for user pysparkbook.
New password:
passwd: all authentication tokens updated successfully.
```

In the preceding code, you can see that the command adduser has been used to create or add a user. The command passwd has been used to provide a password for our new user pysparkbook to the sudo.

```
[root@localhost pyspark]# usermod -aG wheel pyspark
[root@localhost pyspark]#exit
```

Then we will enter to our user pysparkbook.

```
[pyspark@localhost ~]$ su pysparkbook
```

We will create two directories. The binaries directory under the home directory will be used to download software, and the allPySpark directory under the root (/) directory will be used to install big data frameworks:

```
[pysparkbook@localhost ~]$ mkdir binaries
[pysparkbook@localhost ~]$ sudo  mkdir /allPySpark
```

Step 2-1-3. Installing Java

Hadoop, Hive, Spark and many big data frameworks use Java to run on. That's why we are first going to install Java. We are going to use OpenJDK for this purpose; we'll install the eighth version of OpenJDK. We can install Java on CentOS by using the yum installer, as follows:

```
[pysparkbook@localhost binaries]$ sudo yum install java-1.8.0-openjdk.x86_64
```

After installation of any software, it is a good idea to check the installation to ensure that everything is fine.

To check the Java installation, I prefer the command java -version:

[pysparkbook@localhost binaries]$ java -version

The output is as follows:

```
openjdk version "1.8.0_111"
OpenJDK Runtime Environment (build 1.8.0_111-b15)
OpenJDK 64-Bit Server VM (build 25.111-b15, mixed mode)
```

Java has been installed. Now we have to look for the environment variable JAVA_HOME, which will be used by all the distributed frameworks. After installation, JAVA_HOME can be found by using jrunscript as follows:

[pysparkbook@localhostbinaries]$jrunscript -e 'java.lang.System.out.println(java.lang.System.getProperty("java.home"));'

Here is the output:

```
/usr/lib/jvm/java-1.8.0-openjdk-1.8.0.111-2.b15.el7_3.x86_64/jre
```

Step 2-1-4. Creating Passwordless Logging from pysparkbook

Use this command to create a passwordless login:

[pysparkbook@localhost binaries]$ ssh-keygen -t rsa

Here is the output:

```
Generating public/private rsa key pair.
Enter file in which to save the key (/home/pysparkbook/.ssh/id_rsa):
/home/pysparkbook/.ssh/id_rsa already exists.
Overwrite (y/n)? y
Enter passphrase (empty for no passphrase):
Enter same passphrase again:
Your identification has been saved in /home/pysparkbook/.ssh/id_rsa.
Your public key has been saved in /home/pysparkbook/.ssh/id_rsa.pub.
The key fingerprint is:
fd:9a:f3:9d:b6:66:f5:29:9f:b5:a5:bb:34:df:cd:6c pysparkbook@localhost.localdomain
The key's randomart image is:
+--[ RSA 2048]----+
|                 |
|                 |
|                 |
|        .        |
|       S .       |
|        .    .|
```

```
|        . o.=|
|      .o ++OE|
|      oo.+XO*|
+-----------------+
```

```
[pysparkbook@localhost binaries]$ cat ~/.ssh/id_rsa.pub >> ~/.ssh/authorized_keys
[pysparkbook@localhost binaries]$ chmod 755 ~/.ssh/authorized_keys
[pysparkbook@localhost binaries]$ ssh localhost
```

Here is the output:

```
Last login: Wed Dec 21 16:17:45 2016 from localhost
```

```
[pysparkbook@localhost ~]$ exit
```

Here is the output:

```
logout
Connection to localhost closed.
```

Step 2-1-5. Downloading Hadoop

We are going to download Hadoop from the Apache website. As noted previously, we will download all the software into the binaries directory. We'll use the wget command to download Hadoop:

```
[pysparkbook@localhost ~]$ cd  binaries
[pysparkbook@localhost binaries]$ wget http://redrockdigimark.com/
apachemirror/hadoop/common/hadoop-2.6.5/hadoop-2.6.5.tar.gz
```

Here is the output:

```
--2016-12-21 12:50:55--  http://redrockdigimark.com/apachemirror/hadoop/
common/hadoop-2.6.5/hadoop-2.6.5.tar.gz
Resolving redrockdigimark.com (redrockdigimark.com)... 119.18.61.94
Connecting to redrockdigimark.com (redrockdigimark.com)|119.18.61.94|:80...
connected.
HTTP request sent, awaiting response... 200 OK
Length: 199635269 (190M) [application/x-gzip]
Saving to: 'hadoop-2.6.5.tar.gz'
```

Step 2-1-6. Moving Hadoop Binaries to the Installation Directory

Our installation directory is allPySpark. The downloaded software is hadoop-2.6.5.tar.gz, which is a compressed directory. So at first we have to decompress it by using the tar command as follows:

```
[pysparkbook@localhost binaries]$ tar xvzf hadoop-2.6.5.tar.gz
```

Now we'll move Hadoop under the allPySpark directory:

pysparkbook@localhost binaries]$ sudo mv hadoop-2.6.5   /allPySpark/hadoop

Step 2-1-7. Modifying the Hadoop Environment File

We have to make some changes in the Hadoop environment file. This file is found in the Hadoop configuration directory. In our case, the Hadoop configuration directory is /allPySpark/hadoop/etc/hadoop/. Use the following line of code to add JAVA_HOME to the hadoop-env.sh file:

[pysparkbook@localhost binaries]$ vim /allPySpark/hadoop/etc/hadoop/hadoop-env.sh

After opening the Hadoop environment file, add the following line:

```
# The java implementation to use.
export JAVA_HOME=/usr/lib/jvm/java-1.8.0-openjdk-1.8.0.111-2.b15.el7_3.x86_64/jre
```

Step 2-1-8. Modifying the Hadoop Properties Files

In this step, we are concerned with three properties files:

- hdfs-site.xml: HDFS properties

- core-site.xml: Core properties related to the cluster

- mapred-site.xml: Properties for the MapReduce framework

These properties files are found in the Hadoop configuration directory. In the preceding chapter, we discussed HDFS. You learned that HDFS has two components: NameNode and DataNode. You also learned that HDFS uses data replication for fault-tolerance. In our hdfs-site.xml file, we are going to set the NameNode directory by using the dfs.name.dir parameter, the DataNode directory by using the dfs.data.dir parameter, and the replication factor by using the dfs.replication parameter.

Let's modify hdfs-site.xml:

[pysparkbook@localhost binaries]$ vim /allPySpark/hadoop/etc/hadoop/hdfs-site.xml

After opening hdfs-site.xml, we have to put the following lines in that file:

```
<property>
  <name>dfs.name.dir</name>
    <value>file:/allPySpark/hdfs/namenode</value>
    <description>NameNode location</description>
</property>
```

```
<property>
  <name>dfs.data.dir</name>
    <value>file:/allPySpark/hdfs/datanode</value>
      <description>DataNode location</description>
</property>

<property>
 <name>dfs.replication</name>
 <value>1</value>
 <description> Number of block replication </description>
</property>
```

After updating hdfs-site.xml, we are going to update core-site.xml.
In core-site.xml, we are going to update only one property, fs.default.name.
This property is used to determine the host, port, and other details of the file system.
Add the following lines to core-site.xml:

```
<property>
  <name>fs.default.name</name>
    <value>hdfs://localhost:9746</value>
</property>
```

Finally, we are going to modify mapred-site.xml. We also are going to modify
mapreduce.framework.name, which will decide which runtime framework has to be used.
The possible values are local, classic, or yarn. Add the following to mapred-site.xml:

[pysparkbook@localhost binaries]$ cp /allPySpark/hadoop/etc/hadoop/mapred-site.xml.template /allPySpark/hadoop/etc/hadoop/mapred-site.xml

[pysparkbook@localhost binaries]$vim /allPySpark/hadoop/etc/hadoop/mapred-site.xml

```
<property>
 <name>mapreduce.framework.name</name>
   <value>yarn</value>
</property>
```

Step 2-1-9. Updating the .bashrc File

Next, we'll add the following lines to the .bashrc file. Open the .bashrc file:

[pysparkbook@localhost binaries]$ vim ~/.bashrc

Then add the following lines:

```
export HADOOP_HOME=/allPySpark/hadoop
export PATH=$PATH:$HADOOP_HOME/sbin
export PATH=$PATH:$HADOOP_HOME/bin
export JAVA_HOME=/usr/lib/jvm/java-1.8.0-openjdk-1.8.0.111-2.b15.el7_3.x86_64/jre
export PATH=$PATH:$JAVA_HOME/bin
```

Then we have to source the .bashrc file. After sourcing the file, the new updated values will be reflected in the console.

[pysparkbook@localhost binaries]$ source ~/.bashrc

Step 2-1-10. Running the NameNode Format

We have updated some property files. We are supposed to run in NameNode format so that all the changes are reflected in our framework. Use the following command to run NameNode format:

[pysparkbook@localhost binaries]$ hdfs namenode -format

Here is the output:

```
16/12/22 02:38:15 INFO namenode.NameNode: STARTUP_MSG:
/************************************************************
STARTUP_MSG: Starting NameNode
STARTUP_MSG:    host = localhost/127.0.0.1
STARTUP_MSG:    args = [-format]
STARTUP_MSG:    version = 2.6.5
.
16/12/22 02:40:59 INFO util.ExitUtil: Exiting with status 0
16/12/22 02:40:59 INFO namenode.NameNode: SHUTDOWN_MSG:
/************************************************************
SHUTDOWN_MSG: Shutting down NameNode at localhost/127.0.0.1
************************************************************/
```

Step 2-1-11. Starting Hadoop

Hadoop has been installed, and now we can start it. We can find the Hadoop starting script in /allPySpark/hadoop/sbin/. Although this script has been deprecated, we'll use it for this example:

[pysparkbook@localhost binaries]$ /allPySpark/hadoop/sbin/start-all.sh

Step 2-1-12. Checking the Hadoop Installation

We know that the jps command will show all the Java processes running on the machine. Here is the command:

[pysparkbook@localhost binaries]$ jps

If everything is fine, we will see the process running as shown here:

```
25720 NodeManager
25896 Jps
25625 ResourceManager
25195 NameNode
25292 DataNode
25454 SecondaryNameNode
```

Congratulations! We have finally installed Hadoop on our systems.

Recipe 2-2. Install Spark on a Single Machine

Problem

You want to install Spark on a single machine.

Solution

We are going to install the prebuilt spark-2.0.0 for Hadoop version 2.6. We can build Spark from source code. But in this example, we are going to use the prebuilt Apache Spark.

How It Works

Follow the steps in this section to complete the installation.

Step 2-2-1. Downloading Apache Spark

We are going to download Spark from its mirror. We'll use the wget command as follows:

```
[pysparkbook@localhost binaries]$ wget  https://d3kbcqa49mib13.cloudfront.
net/spark-2.0.0-bin-hadoop2.6.tgz
```

Step 2-2-2. Extracting a .tgz file of Spark

Use this command to extract the .tgz file:

[pysparkbook@localhost binaries]$ tar xvzf spark-2.0.0-bin-hadoop2.6.tgz

Step 2-2-3. Moving the Extracted Spark Directory to /allPySpark

Now we have to move the extracted Spark directory under the /allPySpark location. Use this command:

[pysparkbook@localhost binaries]$ sudo mv spark-2.0.0-bin-hadoop2.6 /allPySpark/spark

Step 2-2-4. Changing the Spark Environment File

The Spark environment file possesses all the environment variables required to run Spark. We are going to set the following environment variables in an environment file:

- HADOOP_CONF_DIR: Configuration directory of Hadoop
- SPARK_CONF_DIR: Alternate configuration directory (default: ${SPARK_HOME}/conf)
- SPARK_LOG_DIR: Stores log files (default: ${SPARK_HOME}/log)
- SPARK_WORKER_DIR: Sets the working directory of worker processes
- HIVE_CONF_DIR: Used to read data from Hive

First we have to copy the spark-env.sh.template file to spark-env.sh. The Spark environment file, spark-env.sh, is found inside spark/conf (the configuration directory location). Here is the command:

[pysparkbook@localhost binaries]$ cp /allPySpark/spark/conf/spark-env. sh.template /allPySpark/spark/conf/spark-env.sh

Now let's open the spark-env.sh file:

[pysparkbook@localhost binaries]$ vim /allPySpark/spark/conf/spark-env.sh

Now append the following lines to the end of spark-env.sh:

```
export HADOOP_CONF_DIR=/allPySpark/hadoop/etc/hadoop/
export SPARK_LOG_DIR=/allPySpark/logSpark/
export SPARK_WORKER_DIR=/tmp/spark
export HIVE_CONF_DIR=/allPySpark/hive/conf
```

Step 2-2-5. Amending the .bashrc File

In the .bashrc file, we have to add the Spark bin directory. Use the following command:

[pysparkbook@localhost binaries]$ vim ~/.bashrc

Then add the following lines in the .bashrc file:

```
export SPARK_HOME=/allPySpark/spark
export PATH=$PATH:$SPARK_HOME/bin
```

After this, source the .bashrc file:

[pysparkbook@localhost binaries]$ source ~/.bashrc

Step 2-2-6. Starting PySpark

We can start the PySpark shell by using the pyspark script. Discussion of the pyspark script will continue in the next recipe.

```
[pysparkbook@localhost binaries]$ pyspark
```

We have completed one more successful installation. Other installations are required to move through this book, but first let's focus on the PySpark shell.

Recipe 2-3. Use the PySpark Shell
Problem

You want to use the PySpark shell.

Solution

The PySpark shell is an interactive shell for interacting with PySpark by using Python. The PySpark shell can be started by using a PySpark script. The PySpark script can be found at the spark/bin location.

How It Works

The PySpark shell can be started as follows:

[pysparkbook@localhost binaries]$ pyspark

After starting, PySpark will show the screen in Figure 2-1.

Figure 2-1. *Starting up the console screen in PySpark*

You can see that, after starting, PySpark displays a lot of information. It displays information about the Python version it is using as well as the PySpark version.

The >>> symbol is known to Python programmers. Whenever we start the Python shell, we get this symbol. It tells us that we can now write our Python commands. Similarly, in PySpark, this symbol tells us that now we can write our Python or PySpark commands and see the results.

The PySpark shell works similarly on both a single-machine installation and a cluster installation of PySpark.

Recipe 2-4. Install Hive on a Single Machine

Problem

You want to install Hive on a single machine.

Solution

We discussed Hive in the first chapter. Now it is time to install Hive on our machines. We are going to read data from Hive in PySpark in upcoming chapters.

How It Works

Follow the steps in this section to complete the installation.

Step 2-4-1. Downloading Hive

We can download Hive from the Apache Hive website. We download the Hive tar.gz file by using the wget command as follows:

```
[pysparkbook@localhost binaries]$ wget http://www-eu.apache.org/dist/hive/
hive-2.0.1/apache-hive-2.0.1-bin.tar.gz
```

Here is the output:

```
100%[===========================================>]
139,856,338  709KB/s   in 3m 5s

2016-12-26 09:34:21 (737 KB/s) - 'apache-hive-2.0.1-bin.tar.gz' saved
[139856338/139856338]
```

Step 2-4-2. Extracting Hive

We have downloaded apache-hive-2.0.1-bin.tar.gz, a .tar.gz. So now we have to extract it. We can extract it by using the tar command as follows:

```
[pysparkbook@localhost binaries]$ tar xvzf   apache-hive-2.0.1-bin.tar.gz
```

Step 2-4-3. Moving the Extracted Hive Directory

```
[pysparkbook@localhost binaries]$ sudo mv apache-hive-2.0.1-bin /allPySpark/hive
```

Step 2-4-4. Updating hive-site.xml

Hive is dispatched with the embedded Derby database for metastores. The Derby database is memory-less. Hence it is better to provide a definite location for it. We can provide that location in hive-site.xml. For that, we have to move hive-default.xml. template to hive-site.xml.

[pysparkbook@localhost binaries]$ mv /allPySpark/hive/conf/hive-default.xml.
template /allPySpark/hive/conf/hive-default.xml.templatehive-site.xml

Then open hive-site.xml and update the following:

[pysparkbook@localhost binaries]$ vim /allPySpark/hive/conf/hive-site.xml

You can add the following lines to the end of hive-site.xml or you can change javax.jdo.option.ConnectionURL in the hive-site.xml file:

```
<name>javax.jdo.option.ConnectionURL</name>
    <value>jdbc:derby:;databaseName=/allPySpark/hive/metastore/metastore_db;
create=true</value>
```

After that, we have to add HADOOP_HOME to the hive_env.sh file, as follows:

[pysparkbook@localhost binaries]$ mv /allPySpark/hive/conf/hive-env.
sh.template /allPySpark/hive/conf/hive-env.sh

[pysparkbook@localhost binaries]$ vim /allPySpark/hive/conf/hive-env.sh

And in hive-env.sh, add the following line:

```
# Set HADOOP_HOME to point to a specific hadoop install directory
HADOOP_HOME=/allPySpark/hadoop
```

Step 2-4-5. Updating the .bashrc File

Open the .bashrc file. This file will stay in the home directory:

[pysparkbook@localhost binaries]$ vim ~/.bashrc

Add the following lines into the .bashrc file:

```
####################Hive Parameters #####################
export HIVE_HOME=/allPySpark/hive
export PATH=$PATH:$HIVE_HOME/bin
```

Now source the .bashrc file by using following command:

```
[pysparkbook@localhost binaries]$ source ~/.bashrc
```

Step 2-4-6. Creating Data Warehouse Directories of Hive

Now we have to create data warehouse directories. This data warehouse directory is used by Hive to place the data files.

```
[pysparkbook@localhost binaries]$hadoop fs -mkdir -p /user/hive/warehouse
[pysparkbook@localhost binaries]$hadoop fs -mkdir -p /tmp
[pysparkbook@localhost binaries]$hadoop fs -chmod g+w /user/hive/warehouse
[pysparkbook@localhost binaries]$hadoop fs -chmod g+w /tmp
```

The /user/hive/warehouse directory is the Hive warehouse directory.

Step 2-4-7. Initiating the Metastore Database

Sometimes it is necessary to initiate a schema. You might be thinking, a schema of what? We know that Hive stores metadata of tables in a relational database. For the time being, we are going to use a Derby database as the metastore database for Hive. And then, in upcoming recipes, we are going to connect our Hive to an external PostgreSQL. In Ubuntu, Hive installation works without this command. But in CentOS I found it indispensable to run. Without the following command, Hive throws errors:

```
[pysparkbook@localhost  binaries]$ schematool -initSchema -dbType derby
```

Step 2-4-8. Checking the Hive Installation

Now that Hive has been installed, we should check our work. Start the Hive shell by using the following command:

```
[pysparkbook@localhost binaries]$ hive
```

Then we will find that the Hive shell has been opened as follows:

```
hive>
```

Recipe 2-5. Install PostgreSQL

Problem

You want to install PostgreSQL.

Solution

PostgreSQL is a relational database management system developed at the University of California. The PostgreSQL license provides permission to use, modify, and distribute PostgreSQL. PostgreSQL can run on macOS and on Unix-like systems such as Red Hat and Ubuntu. We are going to install it on CentOS.

We are going to use our PostgreSQL in two ways. First, we'll use PostgreSQL as a metastore database for Hive. After having an external database as a metastore, we will be able to easily read data from the existing Hive. Second, we are going to read data from PostgreSQL, and after analysis, we will save our result to PostgreSQL.

Installing PostgreSQL can be done with source code, but we are going to install it via the command-line yum installer.

How It Works

Follow the steps in this section to complete the installation.

Step 2-5-1. Installing PostgreSQL

PostgreSQL can be installed using the yum installer. Here is the command:

```
[pysparkbook@localhost binaries]$ sudo yum install postgresql-server
```

[sudo] password for pysparkbook:

Step 2-5-2. Initializing the Database

To use PostgreSQL, we first need to use the initdb utility to initialize the database. If we don't initialize the database, we cannot use it. At the time of database initialization, we can also specify the data file of the database. After installing PostgreSQL, we have to initialize it. The database can be initialized using following command:

```
[pysparkbook@localhost binaries]$ sudo postgresql-setup initdb
```

Here is the output:

```
[sudo] password for pysparkbook:
Initializing database ... OK
```

Step 2-5-3. Enabling and Starting the Database

```
[pysparkbook@localhost binaries]$ sudo systemctl enable postgresql
[pysparkbook@localhost binaries]$ sudo systemctl start postgresql

[pysparkbook@localhost binaries]$ sudo -i -u postgres
```

Here is the output:

```
[sudo] password for pysparkbook:
-bash-4.2$ psql
psql (9.2.18)
Type "help" for help.

postgres=#
```

■ **Note**   The installation procedure is located at the following web page:

```
https://wiki.postgresql.org/wiki/YUM_Installation
```

Recipe 2-6. Configure the Hive Metastore on PostgreSQL

Problem

You want to configure the Hive metastore on PostgreSQL.

Solution

As we know, Hive puts metadata of tables in a relational database. We have already installed Hive. Our Hive installation has an embedded metastore. Hive uses the Derby relational database system for its metastore. In upcoming chapters, we will read existing Hive tables from PySpark.

Configuring the Hive metastore on PostgreSQL requires us to populate tables in the PostgreSQL database. These tables will hold the metadata of the Hive tables. After this, we have to configure the Hive property file.

How It Works

In this section, we are going to configure the Hive metastore on the PostgreSQL database. Then our Hive will have metadata in PostgreSQL.

Step 2-6-1. Downloading the PostgreSQL JDBC Connector

We need a JDBC connector so that the Hive process can connect to the external PostgreSQL. We can get a JDBC connector by using the following command:

```
[pysparkbook@localhost binaries]$ wget https://jdbc.postgresql.org/download/
postgresql-9.4.1212.jre6.jar
```

Step 2-6-2. Copying the JDBC Connector to the Hive lib Directory

After getting the JDBC connector, we have to put it in the Hive lib directory:

```
[pysparkbook@localhost binaries]$ cp postgresql-9.4.1212.jre6.jar
/allPySpark/hive/lib/
```

Step 2-6-3. Connecting to PostgreSQL

Use this command to connect to PostgreSQL:

```
[pysparkbook@localhost binaries]$ sudo -u postgres psql
```

Step 2-6-4. Creating the Required User and Database

In this step, we are going to create a PostgreSQL user, pysparkBookUser. Then we are going to create a database named pymetastore. Our database is going to hold all the tables related to the Hive metastore.

First, create the user:

```
postgres=# CREATE USER pysparkBookUser WITH PASSWORD 'pbook';
```

Here is the output:

```
CREATE ROLE
```

Next, create the database:

```
postgres=# CREATE DATABASE pymetastore;
```

Here is the output:

```
CREATE DATABASE
```

The \c PostgreSQL command stands for *connect*. We have created our database pymetastore. Now we are going to connect to this database by using our \c command:

postgres=# \c pymetastore;

You are now connected to the pymetastore database. You can see more PostgreSQL commands at www.postgresql.org/docs/9.0/static/app-psql.html.

Step 2-6-5. Populating Data in the pymetastore Database

Hive possess its own PostgreSQL scripts to populate tables for the metastore. The \i command reads commands from the PostgreSQL script and executes those commands. The following command runs the hive-txn-schema-2.0.0.postgres.sql script, which will create all the tables required for the Hive metastore:

pymetastore=# \i /allPySpark/hive/scripts/metastore/upgrade/postgres/ hive-txn-schema-2.0.0.postgres.sql

Here is the output:

```
psql:/allPySpark/hive/scripts/metastore/upgrade/postgres/hive-txn-schema-
2.0.0.postgres.sql:30: NOTICE:  CREATE TABLE / PRIMARY KEY will create
implicit index "txns_pkey" for table "txns"
CREATE TABLE
CREATE TABLE
INSERT 0 1
psql:/allPySpark/hive/scripts/metastore/upgrade/postgres/hive-txn-schema-
2.0.0.postgres.sql:69: NOTICE:  CREATE TABLE / PRIMARY KEY will create
implicit index "hive_locks_pkey" for table "hive_locks"
CREATE TABLE
```

Step 2-6-6. Granting Permissions

The following commands will grant some permissions:

pymetastore=# grant select, insert,update,delete on public.txns to pysparkBookUser;
GRANT

pymetastore=# grant select, insert,update,delete on public.txn_components to pysparkBookUser;
GRANT

```
pymetastore=# grant select, insert,update,delete on public.completed_txn_
components    to pysparkBookUser;
GRANT

pymetastore=# grant select, insert,update,delete on public.next_txn_id to
pysparkBookUser;
GRANT

pymetastore=# grant select, insert,update,delete on public.hive_locks to
pysparkBookUser;
GRANT

pymetastore=# grant select, insert,update,delete on public.next_lock_id to
pysparkBookUser;
GRANT

pymetastore=# grant select, insert,update,delete on public.compaction_queue
to pysparkBookUser;
GRANT

pymetastore=# grant select, insert,update,delete on public.next_compaction_
queue_id to pysparkBookUser;
GRANT

pymetastore=# grant select, insert,update,delete on public.completed_
compactions to pysparkBookUser;
GRANT

pymetastore=# grant select, insert,update,delete on public.aux_table to
pysparkBookUser;
GRANT
```

Step 2-6-7. Changing the pg_hba.conf File

Remember that in order to update pg_hba.conf, you are supposed to be the root user. So first become the root user. Then open the pg_hba.conf file:

[root@localhost binaries]# vim /var/lib/pgsql/data/pg_hba.conf

Then change all the peers and indent them to trust:

```
#local   all              all                                     peer
local    all              all                                     trust
# IPv4 local connections:
#host    all              all             127.0.0.1/32            ident
host     all              all             127.0.0.1/32            trust
```

```
# IPv6 local connections:
#host    all            all            ::1/128                ident
host     all            all            ::1/128                trust
```

You can find more details about this change at http://stackoverflow.com/questions/2942485/psql-fatal-ident-authentication-failed-for-user-postgres. Come out of root user.

Step 2-6-8. Testing Our User

Next, we'll test that we are easily able to enter our database as our created user:

[pysparkbook@localhost binaries]$ psql -h localhost -U pysparkbookuser -d pymetastore

Here is the output:

```
psql (9.2.18)
Type "help" for help.
pymetastore=>
```

Step 2-6-9. Modifying Our hive-site.xml

We can modify the Hive-related configuration in the configuration file hive-site.xml. We have to modify the following properties:

- javax.jdo.option.ConnectionURL: Connection URL to the database

- javax.jdo.option.ConnectionDriverName: Connection JDBC driver name

- javax.jdo.option.ConnectionUserName: Database connection user

- javax.jdo.option.ConnectionPassword: Connection password

Either modify these properties or add the following lines at the end of the Hive property file to get the required result:

```
<property>
    <name>javax.jdo.option.ConnectionURL</name>
    <value>jdbc:postgresql://localhost/pymetastore</value>
    <description>postgreSQL server metadata store</description>
</property>
<property>
    <name>javax.jdo.option.ConnectionDriverName</name>
    <value>org.postgresql.Driver</value>
    <description>Driver class of postgreSQL</description>
</property>
```

35

```
  <property>
      <name>javax.jdo.option.ConnectionUserName</name>
      <value>pysparkbookuser</value>
      <description>User name to connect to postgreSQL</description>
  </property>
  <property>
      <name>javax.jdo.option.ConnectionPassword</name>
      <value>pbook</value>
      <description>password for connecting to PostgreSQL server</
description>
  </property>
```

Step 2-6-10. Starting Hive

We have connected Hive to an external relational database management system. So now it is time to start Hive and check that everything is fine. First, start Hive:

[pysparkbook@localhost binaries]$ hive

Our activities will be reflected in PostgreSQL. Let's create a database and a table inside the database. We'll create the database apress and the table apressBooks via the following commands:

hive> create database apress;

Here is the output:

```
OK
Time taken: 1.397 seconds
```

hive> use apress;

Here is the output:

```
OK
Time taken: 0.07 seconds
```

hive> create table apressBooks (
**   >   bookName String,**
**   >   bookWriter String**
**   >   )**
**   >   row format delimited**
**   >   fields terminated by ',';**

Here is the output:

```
OK
Time taken: 0.581 seconds
```

Step 2-6-11. Testing Creation of Metadata in PostgreSQL

The created database and table will be reflected in PostgreSQL. We can see the updated data in the TBLS table as follows:

pymetastore=> SELECT * from "TBLS";

```
TBL_ID | CREATE_TIME | DB_ID | LAST_ACCESS_TIME |    OWNER    | RETENTION |
SD_ID |   TBL_NAME   |
   TBL_TYPE   | VIEW_EXPANDED_TEXT | VIEW_ORIGINAL_TEXT
--------+-------------+-------+------------------+-------------+-----------
+-------+-------------+
---------------+--------------------+--------------------
     1 |  1482892229 |   6 |                0 | pysparkbook |           0
|    1 | apressbooks |
 MANAGED_TABLE |                    |
(1 row)
```

The significant work needed to connect Hive to an external database is done. In the following recipe, we are going to install Apache Mesos.

Recipe 2-7. Connect PySpark to Hive

Problem

You want to connect PySpark to Hive.

Solution

PySpark needs the Hive property file to know the configuration parameters of Hive. The Hive property file, hive-site.xml, stays in the Hive configuration directory. Copy the Hive property file to the Spark configuration directory. Then we will be finished and we can start PySpark.

How It Works

Two steps have been identified to connect PySpark to Hive.

Step 2-7-1. Copying the Hive Property File to the Spark Conf Directory

Use this command to copy the Hive property file:

```
[pysparkbook@localhost binaries]$cp /allPySpark/hive/conf/hive-site.xml /
allPySpark/spark/
```

Step 2-7-2. Starting PySpark

Use this command to start PySpark:

```
[pysparkbook@localhost binaries]$pyspark
```

Recipe 2-8. Install Apache Mesos

Problem

You want to install Apache Mesos.

Solution

Installing Apache Mesos requires downloading the code and then configuring it.

How It Works

Follow the steps in this section to complete the installation.

Step 2-8-1. Downloading Apache Mesos

Use this command to obtain Apache Mesos:

```
[pysparkbook@localhost binaries]$ wget http://www.apache.org/dist/
mesos/1.1.0/mesos-1.1.0.tar.gz
```

The output is as follows:

```
--2016-12-28 08:15:14-- http://www.apache.org/dist/mesos/1.1.0/mesos-
1.1.0.tar.gz
Resolving www.apache.org (www.apache.org)... 88.198.26.2, 140.211.11.105,
2a01:4f8:130:2192::2
Connecting to www.apache.org (www.apache.org)|88.198.26.2|:80... connected.
HTTP request sent, awaiting response... 200 OK
Length: 41929556 (40M) [application/x-gzip]
Saving to: 'mesos-1.1.0.tar.gz.1'
```

```
100%[=============================================>]
41,929,556    226KB/s    in 58s

2016-12-28 08:16:23 (703 KB/s) - 'mesos-1.1.0.tar.gz.1' saved
[41929556/41929556]
```

Step 2-8-2. Extracting Mesos from .tar.gz

To extract Mesos, use this command:

```
[pysparkbook@localhost binaries]$ tar xvzf mesos-1.1.0.tar.gz
```

Step 2-8-3. Installing Repo to Install Maven

To install Maven, we first need to install the repo:

```
[pysparkbook@localhost binaries]$ sudo bash -c 'cat > /etc/yum.repos.d/
wandisco-svn.repo <<EOF
> [WANdiscoSVN]
> name=WANdisco SVN Repo 1.9
> enabled=1
> baseurl=http://opensource.wandisco.com/centos/7/svn-1.9/RPMS/$basearch/
> gpgcheck=1
> gpgkey=http://opensource.wandisco.com/RPM-GPG-KEY-WANdisco
> EOF'
```

Step 2-8-4. Installing Dependencies of Maven

It is time to install the dependencies required to install Maven:

```
[pysparkbook@localhost binaries]$ sudo yum install -y apache-maven python-
devel java-1.8.0-openjdk-devel zlib-devel libcurl-devel openssl-devel cyrus-
sasl-devel cyrus-sasl-md5 apr-devel subversion-devel apr-util-devel
```

Step 2-8-5. Downloading Apache Maven

Now we're ready to download Maven:

```
[pysparkbook@localhost binaries]$ wget http://www-us.apache.org/dist/maven/
maven-3/3.3.9/binaries/apache-maven-3.3.9-bin.tar.gz
```

Here is the output:

```
100%[====================================================>]
8,491,533    274KB/s    in 87s

2016-12-28 23:47:40 (95.5 KB/s) - 'apache-maven-3.3.9-bin.tar.gz' saved
[8491533/8491533]
```

Step 2-8-6. Extracting the Maven Directory

As with other software, we have to extract and move the Maven directory:

[pysparkbook@localhost binaries]$ tar -xvzf apache-maven-3.3.9-bin.tar.gz
[pysparkbook@localhost binaries]$ mv apache-maven-3.3.9 /allPySpark/maven

We then have to link the mvn file:

**[pysparkbook@localhost binaries]$ sudo ln -s /allPySpark/maven/bin/mvn /usr/
bin/mvn**

[sudo] password for pysparkbook:

Step 2-8-7. Checking the Maven Installation

The best way to check our installation is to run the version command:

[pysparkbook@localhost binaries]$ mvn -version

Here is the output:

```
Apache Maven 3.3.9 (bb52d8502b132ec0a5a3f4c09453c07478323dc5; 2015-11-
10T22:11:47+05:30)
Maven home: /allPySpark/maven
Java version: 1.8.0_111, vendor: Oracle Corporation
Java home: /usr/lib/jvm/java-1.8.0-openjdk-1.8.0.111-2.b15.el7_3.x86_64/jre
Default locale: en_US, platform encoding: UTF-8
OS name: "linux", version: "3.10.0-514.2.2.el7.x86_64", arch: "amd64",
family: "unix"
```

Step 2-8-9. Configuring Mesos

We have to move to the Mesos build directory. Then we have to run configure for the script.

[pysparkbook@localhost build]$ ../configure

```
------------------------------------------------------------------
                 Following lines will come as notes
------------------------------------------------------------------
```

You may get errors as follows:

```
make[2]: *** [../3rdparty/protobuf-2.6.1/python/dist/protobuf-2.6.1-
py2.7.egg] Error 1
make[2]: Leaving directory '/allPySpark/mesos/build/src'
make[1]: *** [all] Error 2
make[1]: Leaving directory '/allPySpark/mesos/build/src'
make: *** [all-recursive] Error 1
```

If you do get these errors, you need to perform the following installations and upgrade pytz:

```
[pysparkbook@localhost build]$sudo yum install python-setuptools python-
setuptools-devel
[pysparkbook@localhost build]$sudo easy_install pip
[pysparkbook@localhost build]$sudo pip install --upgrade pytz
------------------------------------------------------------------
```

Step 2-8-10. Running Make

```
[pysparkbook@localhost build]$make
Installing build/bdist.linux-x86_64/wheel/mesos.scheduler-1.1.0-py2.7-nspkg.
pth
running install_scripts
creating build/bdist.linux-x86_64/wheel/mesos.scheduler-1.1.0.dist-info/
WHEEL
make[2]: Leaving directory '/allPySpark/mesos/build/src'
make[1]: Leaving directory '/allPySpark/mesos/build/src'
```

Step 2-8-11. Running make install

Use this command to install Mesos:

```
[pysparkbook@localhost build]$make install
```

Step 2-8-12. Starting Mesos Master

After successful installation of Mesos, we can start the master by using the following command:

```
[pysparkbook@localhost build]$ mesos-master --work_dir=/allPySpark/mesos/workdir
```

Step 2-8-13. Starting Mesos Slaves

Use this command to start the slave on the same machine:

```
[root@localhost binaries]#mesos-slave --master=127.0.0.1:5050 --work_dir=/
allPySpark/mesos/workdir --systemd_runtime_directory=/allPySpark/mesos/systemd
```

In an upcoming chapter, you will see how to start the PySpark shell on Mesos.

Recipe 2-9. Install HBase

Problem

You want to install Apache HBase.

Solution

HBase is a NoSQL database, as we discussed in Chapter 1. We are going to install HBase. Then we will read data from this HBase installation by using Spark. We will go for the simplest installation of HBase. We can download HBase from the HBase website (https://hbase.apache.org/) and then configure it.

How It Works

Follow the steps in this section to complete the installation.

Step 2-9-1. Obtaining HBase

Use the following command to download HBase in our binaries directory:

```
[pysparkbook@localhost binaries]$ wget http://www-eu.apache.org/dist/hbase/
stable/hbase-1.2.4-bin.tar.gz
```

Step 2-9-2. Extracting HBase

```
[pysparkbook@localhost binaries]$ tar xzf  hbase-1.2.4-bin.tar.gz
[pysparkbook@localhost binaries]$ sudo mv hbase-1.2.4 /usr/local/hbase
```

Step 2-9-3. Updating the HBase Environment File

HBase also looks for JAVA_HOME. So we'll update the HBase environment file with JAVA_HOME:

```
[pysparkbook@localhost binaries]$ vim /allPySpark/hbase/conf/hbase-env.sh
```

```
export JAVA_HOME=/usr/lib/jvm/java-1.8.0-openjdk-1.8.0.111-2.b15.el7_3.x86_64/
```

Step 2-9-4. Creating the HBase Directory in HDFS

Even HBase can put its data on a local machine, in the case of a single-machine installation. But to make things clearer about the workings of HBase, we are going to create a directory on HDFS for HBase. First, we have to start HDFS. Use the following commands to create a directory in HDFS:

```
[pysparkbook@localhost binaries]$/allPySpark/hadoop/sbin/start-dfs.sh
[pysparkbook@localhost binaries]$/allPySpark/hadoop/sbin/start-yarn.sh
[pysparkbook@localhost binaries]$hadoop fs -mkdir /hbase
```

Step 2-9-5. Updating the HBase Property File and .bashrc

Let's start with updating the property file, and then we will update the .bashrc file. In the HBase property file, we are going to update hbase:rootdir. The HBase property file stays in the HBase configuration directory. For us, the HBase configuration directory is /allPySpark/hbase/conf.

```
[pysparkbook@localhost binaries]$ vim /allPySpark/hbase/conf/hbase-site.xml
```

Now add the following lines in the hbase-site.xml file:

```
<property>
<name>hbase:rootdir</name>
<value>hdfs://localhost:9746/hbase</value>
</property>
```

It is time to update .bashrac, as shown here:

```
[pysparkbook@localhost binaries]$vim ~/.bashrc
```

Add the following lines in .bashrc:

```
export HBASE_HOME=/allPySpark/hbase
export PATH=$PATH:$HBASE_HOME/bin
[pysparkbook@localhost binaries]$ source ~/.bashrc
```

Step 2-9-6. Starting HBase and the HBase Shell

```
[pysparkbook@localhost binaries]$ /allPySpark/hbase/bin/start-hbase.sh
[pysparkbook@localhost binaries]$/allPySpark/hbase/bin/hbase shell
```

Here is the output:

```
hbase(main):001:0>
```

CHAPTER 3

■ ■ ■

Introduction to Python and NumPy

Python is a general-purpose, high-level programming language. It was developed by Guido van Rossum, and since its inception, its popularity has increased exponentially. A plethora of handy and high-performance packages for numerical and statistical calculations make Python popular among data scientists. Python is an indented language, which may bother programmers just learning Python. But indented code improves code readability, and the popularity of Python and its ease of use makes it good for programming Spark.

This chapter introduces the basics of Python programming. We will also discuss NumPy. In addition, I have included a recipe on IPython and on integrating IPython with PySpark.

If you know Python well, you can skip this chapter. But my suggestion is to go through the content regardless., because this chapter might still provide good insight and boost your knowledge further.

This chapter covers the following recipes:

Recipe 3-1. Create data and verify the data type

Recipe 3-2. Create and index a Python string

Recipe 3-3. Typecast from one data type to another

Recipe 3-4. Work with a Python list

Recipe 3-5. Work with a Python tuple

Recipe 3-6. Work with a Python set

Recipe 3-7. Work with a Python dictionary

Recipe 3-8. Work with Define and Call functions

Recipe 3-9. Work with Create and Call lambda functions

Recipe 3-10. Work with Python conditionals

Recipe 3-11. Work with Python for and while loops

© Raju Kumar Mishra 2018
R. K. Mishra, *PySpark Recipes*, https://doi.org/10.1007/978-1-4842-3141-8_3

Recipe 3-1. Create Data and Verify the Data Type

Problem

You want to create data and verify the data type.

Solution

Python is dynamically typed language. What does that mean? *Dynamically typed* means that at the time of variable definition, the programmer is not supposed to mention data types, as we do in other programming languages such as C. To learn about Python data types, you want to do the following:

- Create an integer and verify its data type
- Create a long integer and verify its data type
- Create a decimal and verify its data type
- Create a Boolean and verify its data type

The Python interpreter interprets the data type whenever a literal is mentioned in the console or in a Python script. The Python interpreter interprets the data type of a literal when it is assigned to a variable. In order to verify the data type of a literal or variable, you can use the Python type() function.

How It Works

Let's follow the steps in this section to create data and verify data types.

Step 3-1-1. Creating an Integer and Verifying Its Data Type

Let's create an integer in Python. In following line of code, we associate the Python variable pythonInt with the value 15:

```
>>> pythonInt = 15
>>> pythonInt
```

Here is the output:

```
15
```

We can verify the type of a Python object by using the type() function:

```
>>> type(pythonInt)
```

Here is the output:

```
<type 'int'>
```

Step 3-1-2. Creating a Long Integer and Verifying Its Data Type

The Long data type is used for large integers. At the time of creation, the number is suffixed by L. Creation of a Long data type is shown here:

```
>>> pythonLongInt = 15L
>>> pythonLongInt
```

Here is the output:

```
15L
```

Using the type() function, we can see that data type of pythonLongInt is long.

```
>>> type(pythonLongInt)
```

Here is the output:

```
<type 'long'>
```

Step 3-1-3. Creating a Decimal and Verifying Its Data Type

Decimal numbers are used a lot in any numerical computation. Let's create a floating-point number:

```
>>> pythonFloat = 15.4
>>> pythonFloat
```

Here is the output:

```
15.4
```

And now let's check its type:

```
>>> type(pythonFloat)
```

Here is the output:

```
<type 'float'>
```

Step 3-1-4. Creating a Boolean and Verifying Its Data Type

A programmer's life is filled with concerns about various conditions, such as whether a given number is greater than five. Here's an example:

```
>>> pythonBool = True
>>> pythonBool
```

The output is shown here:

```
True
```

```
>>> type(pythonBool)
```

Here is the output:

```
<type 'bool'>
>>> pythonBool = False
>>> pythonBool
```

Here is the output:

```
False
```

Recipe 3-2. Create and Index a Python String
Problem

You want to create and index a Python string.

Solution

Natural language processing and other string-related problems. You want to do the following:

- Create a string and verify its data type
- Index a string
- Verify whether a substring lies in a given string
- Check whether a string starts with a given substring
- Check whether a string ends with a given substring

You can create a string by using either a set of double quotes (" ") or a set of single quotes (' '). Indexing can be done with a set of square brackets. You can use various ways to verify whether a substring is in a given string. The find() function can be used to find whether a substring will stay inside a string.

The startswith() function indicates whether a given string starts with a given substring. Similarly, the endswith() function can confirm whether a given string ends with given substring.

How It Works

The steps in this section will solve our problem.

Step 3-2-1. Creating a String and Verifying Its Data Type

To create a string, you must put the given string literal between either double quotes or single quotes. Always remember to not use mixed quotes, meaning having a single quote on one side and a double quote on the other side. The following lines of code create strings by using double and single quotes. Let's start with double quotes:

```
>>> pythonString  = "PySpark Recipes"
>>> pythonString
```

Here is the output:

```
'PySpark Recipes'
```

You can use the following code to verifying the type:

```
>>> type(pythonString)
```

Here is the output:

```
<type 'str'>
```

Now let's create a string by using single quotes:

```
>>> pythonString  = 'PySpark Recipes'
>>> pythonString
```

Here is the output:

```
'PySpark Recipes'
```

Step 3-2-2. Indexing a String

String elements can be indexed. When you index a string, you start from zero:

```
>>> pythonStr = "Learning PySpark is fun"
>>> pythonStr
```

Here is the output:

```
'Learning PySpark is fun'
```

Let's get the string at the tenth location. Let me mention again that string indexes start with zero:

```
>>> pythonStr[9]
```

Here is the output:

```
'p'
```

Step 3-2-3. Verifying That a Substring Lies in a Given String

If a substring is found inside a string, the Python string find() function will return the lowest index of the matching substring in a string. But if the given substring is not found in a given string, the find() method returns –1. Let's index the element at the ninth position:

```
>>> pythonStr[9]
```

Here is the output:

```
'p'
```

We are going to search to see whether the substring Py is in our string pythonStr. We'll use the find() method:

```
>>> pythonStr.find("Py")
```

Here is the output:

```
9
```

```
>>> pythonStr.find("py")
```

Here is the output:

```
-1
```

We can see that the output is 9. This is the index where Py is started.

Step 3-2-4. Checking Whether a String Starts with a Given Substring

Now let's focus on another important function, startswith(). This function ensures that a given string starts with a particular substring. If a given string starts with the mentioned substring, then it returns True; otherwise, it returns False. Here is an example:

```
>>> pythonStr.startswith("Learning")
```

Here is the output:

```
True
```

Step 3-2-5. Checking Whether a String Ends with a Given Substring

Similarly, the endswith() function indicates whether a given string ends with given substring.

```
>>> pythonStr.endswith("fun")
```

Here is the output:

```
True
```

Recipe 3-3. Typecast from One Data Type to Another
Problem

You want to typecast from one data type to another.

Solution

Typecasting from one data type to another is a general activity in data analysis. The following set of problems will help you to understand typecasting in Python. You want to do the following:

- Typecast an integer number to a float

- Typecast a string to an integer

- Typecast a string to a float

Typecasting means changing one data type to another—for example, changing a string to an integer or float. In PySpark, we'll often typecast one data type to another. There are four important functions in Python for typecasting one data type to another: int(), float(), long(), and str().

How It Works

We'll follow the steps in this section to solve the given problems.

Step 3-3-1. Typecasting an Integer Number to a Float

Let's start with creating an integer:

```
>>> pythonInt = 17
```

We have created an integer variable, pythonInt. The type of a Python object can be found by using the type() function:

```
>>> type(pythonInt)
```

Here is the output:

```
<type 'int'>
```

We can see clearly that the type() function has returned int.
To typecast any data type to float, we use the float() function:

```
>>> pythonFloat = float(pythonInt)
```

The pythonInt value has been changed to float. We can see the change by printing the variable:

```
>>> print pythonFloat
```

Here is the output:

```
17.0
```

Performing the type() function on pythonFloat will ensure that the integer value has been typecasted to a floating number :

```
>>> type(pythonFloat)
```

Here is the output:

```
<type 'float'>
```

Step 3-3-2. Typecasting a String to an Integer

The function int() typecasts a String, Float, or Long type to an Integer. Typecasting a string to an integer will come up again and again in our problem-solving by PySpark. We'll start by creating a string:

```
>>> pythonString = "15"
>>> type(pythonString)
```

Here is the output:

```
<type 'str'>
```

In the preceding Python code snippet, we have created a string, and we also checked the type of our created variable pythonString.

The Python built-in int() function can typecast a variable to an integer. The following code uses an int() function and changes a string to an integer:

```
>>> pythonInteger = int(pythonString)
>>> pythonInteger
```

Here is the output:

```
15
>>> type(pythonInteger)
```

Here is the output:

```
<type 'int'>
```

Step 3-3-3. Typecasting a String to a Float

Let's solve our last question of typecasting. Let's create a string with the value 15.4, and typecast it to a float by using the float() function:

```
>>> pythonString = "15.4"
>>> type(pythonString)
```

Here is the output:

```
<type 'str'>
```

Next, we'll typecast our string to a floating-point number:

```
>>> pythonFloat = float(pythonString)
>>> pythonFloat
```

Here is the output:

```
15.4
```

```
>>> type(pythonFloat)
```

Here is the output:

```
<type 'float'>
```

There are four types of collections in Python. The four types are list, set, dictionary, and tuple. In the following recipes, we are going to discuss these collections one by one.

Recipe 3-4. Work with a Python List
Problem

You want to work with a Python list

Solution

A *list* is an ordered Python collection. A list is used in many problems. We are going to concentrate on the following problems:

- Creating a list
- Extending a list
- Appending a list
- Counting the number of elements in a list
- Sorting a list

A Python list is mutable. Generally, we create a list with objects of a similar type. But a list can be created of different object types. A list is created using square brackets, [ ]. A List object has many built-in functions to work with. Extending a list can be done by using the extend() function. Appending a list can be done using the append() function. You might be wondering, what is the difference between appending a list and extending a list? Soon you are going to get the answer.

Counting the number of elements in a list can be done by using the len() function, and sorting by using the sort() function.

How It Works

Use the following steps to solve our problem.

Step 3-4-1. Creating a List

Let's create a list:

```
>>> pythonList = [2.3,3.4,4.3,2.4,2.3,4.0]
>>> pythonList
```

Here is the output:

```
[2.3, 3.4, 4.3, 2.4, 2.3, 4.0]
```

A list can be indexed by using square brackets, [ ]. The first element of a list is indexed with 0:

```
>>> pythonList[0]
```

Here is the output:

```
2.3
```

```
>>> pythonList[1]
```

Here is the output:

```
3.4
```

pythonList[0] indexes the first element of the list pythonList, and pythonList[1] indexes the second element. Hence if a list has *n* elements, the last element can be indexed as pythonList[n-1].

A list of elements can be of a different data type. The following example will make it clear:

```
>>> pythonList1 = ["a",1]
```

The preceding line creates a list with two elements. The first element of the list is a string, and the second element is an integer.

```
>>> pythonList1
```

Here is the output:

```
['a', 1]
```

```
>>> type(pythonList1)
```

Here is the output:

```
<type 'list'>
```

The type() function outputs the type of pythonList1 as list:

```
>>> type (pythonList1[0])
```

Here is the output:

```
<type 'str'>
```

```
>>> type (pythonList1[1])
```

Here is the output:

```
<type 'int'>
```

The type of the first element is a string, and the second element is an integer, which is being shown by the type() function.

A list is a mutable collection in Python. A *mutable collection* means the elements of the collection can be changed. Let's look at some examples:

```
>>>pythonList1 = ["a",1]
>>> pythonList1[0] = 5
>>> print  pythonList1
```

Here is the output:

```
[5, 1]
```

In this example, we used the same list, pythonList1, which we created. Then the first element is changed to 5. And we can see by printing pythonList1 that the first element of pythonList1 is 5 now.

Step 3-4-2. Extending a List

The extend() function takes an object as an argument and extends the calling list object with object in argument element wise:

```
>>>pythonList1 = [5,1]
>>>print  pythonList1
```

Here is the output:

```
[5, 1]
```

```
>>> pythonList1.extend([1,2,3])
>>> pythonList1
[5, 1, 1, 2, 3]
```

Step 3-4-3. Appending a List

Applying append() to a list will append the Python object that has been provided as an argument to the function. In this example, append() has just appended another list to the existing list:

```
>>>pythonList1 = [5,1]
>>> pythonList1.append([1,2,3])

>>> print  pythonList1
```

Here is the output:

```
[5, 1, [1, 2, 3]]
```

Step 3-4-4. Counting the Number of Elements in a List

The *length* of a list is the number of elements in the list. The len() function will return the length of a list as follows:

```
>>>pythonList1 = [5,1]
>>> len(pythonList1)
```

Output:

```
2
```

Step 3-4-5. Sorting a List

Sorting a list can be done in an increasing or decreasing fashion. Our sort() function can be applied in the following ways to sort a given list in either ascending or descending order.

Let's start with sorting our list pythonList in an ascending order in the following code:

```
>>> pythonList = [2.3,3.4,4.3,2.4,2.3,4.0]
>>> pythonList
```

Here is the output:

```
[2.3, 3.4, 4.3, 2.4, 2.3, 4.0]
```

```
>>> pythonList.sort()
>>> pythonList
```

Here is the output:

```
[2.3, 2.3, 2.4, 3.4, 4.0, 4.3]
```

In order to sort data in descending order, we have to provide the reverse argument as True:

```
>>> pythonList.sort(reverse=True)
>>> pythonList
```

Here is the output:

```
[4.3, 4.0, 3.4, 2.4, 2.3, 2.3]
```

Recipe 3-5. Work with a Python Tuple
Problem

You want to work with a Python tuple.

Solution

A *tuple* is an immutable ordered collection. We are going to solve the following set of problems:

- Creating a tuple

- Getting the index of an element of a tuple

- Counting the occurrence of a tuple element

A tuple, an immutable collection, is generally used to create record data. A tuple is created by using a set of parentheses, ( ). A tuple is an ordered sequence. We can put different data types together in a tuple. The index( ) function on a tuple object will provide us the index of the first occurrence of a given element. Another function, count( ), defined on a tuple will return the frequency of a given element.

How It Works

Let's work out the solution step-by-step.

Step 3-5-1. Creating a Tuple

The following code creates a tuple by using parentheses, ( ):

```
>>>pythonTuple = (2.0,9,"a",True,"a")
```

Here we have created a tuple, pythonTuple, which has five elements. The first element of our tuple is a decimal number, the second element is an integer, the third one is a string, the fourth one is a Boolean, and last one is a string.

Now let's check the type of the pythonTuple object:

```
>>> type(pythonTuple)
```

Here is the output:

```
<type 'tuple'>
```

Indexing a tuple is done in a similar way as we indexed the list, but this time we use square brackets, [ ]:

```
>>> pythonTuple[2]
```

Here is the output:

```
'a'
```

This next line of code will show that the tuple is immutable. We are

```
>>> pythonTuple[1] = 5
```

Here is the output:

```
Traceback (most recent call last):
  File "<stdin>", line 1, in <module>
TypeError: 'tuple' object does not support item assignment
```

We can see that we cannot modify the elements of a tuple.

Step 3-5-2. Getting the Index of a Tuple Element

The index() function requires an element as an argument and returns the index of the first occurrence of a value in a given tuple. In our tuple pythonTuple, 'a' is at index 2. We can get this index as follows:

```
>>> pythonTuple.index('a')
```

Here is the output:

```
2
```

If a value is not found in a given tuple, the index() function throws an exception.

Step 3-5-3. Counting the Occurrences of a Tuple Element

Providing an element as input in a count() function returns the frequency of that element in a given tuple. We can use the count() function in the following way:

```
>>> pythonTuple.count("a")
```

Here is the output:

```
2
```

In our tuple pythonTuple, the character a has occurred twice. Therefore, the count() function has returned 2.

You can try a simple exercise: apply count() on that tuple with 1 as an argument . You will get 1 as the answer.

Recipe 3-6. Work with a Python Set
Problem

You want to work with a Python set.

Solution

Dealing with a collection of distinct elements requires us to use a set. We are going to work on the following tasks:

- Creating a set

- Adding a new element to a set

- Performing a union on sets

- Performing an intersection operation on sets

A set cannot have duplicate elements. A Python set is created using set of curly brackets, { }. The most important point to note is that, at the time of creation, we can put duplicate items in a set. But the set will then remove all the duplicate items from it. As with other collections, many functions have been defined on the set object too. If you want to add a new element to a set, you can use the add() function.

Unionizing two sets is a common activity that we'll find in our day-to-day tasks. The union() function, which has been defined on our set object, will unionize two sets for us.

The intersect() function is used to run an intersection on two given Python sets.

How It Works

In this section, we will solve our given problems step-by-step.

Step 3-6-1. Creating a Set

Let's create a set of stationary items and then verify the existence of distinct elements. This code line creates a set of stationary items:

```
>>> pythonSet = {'Book','Pen','NoteBook','Pencil','Book'}
>>> pythonSet
```

Here is the output:

```
set(['Pencil', 'Pen', 'Book', 'NoteBook'])
```

We have created a set. We can observe that putting in a duplicate element doesn't throw an error, but the set will not consider that duplicate element while creating the set.

Step 3-6-2. Adding a New Element to a Set

By using the add() function, we can add a new element to the set. We have already created the set pythonSet. Now we are going to add a new element, Eraser, to our set, as shown here:

```
>>> pythonSet.add("Eraser")
>>> pythonSet
```

Here is the output:

```
set(['Pencil', 'Pen', 'Book', 'Eraser', 'NoteBook'])
```

You can see in this example that we have added the new element Eraser to our set.

Step 3-6-3. Performing a Union on Sets

A union operation on a Python set will behave in a mathematical way. Let's create another set for this example:

```
>>> pythonSet1 = {'NoteBook','Pencil','Diary','Marker'}
>>> pythonSet1
```

Here is the output:

```
set(['Marker', 'Pencil', 'NoteBook', 'Diary'])
```

A union of two sets will return another set with all the elements either in any set or common to both sets. In the following example, we can see that the union of pythonSet and pythonSet1 has returned all the merged elements in pythonSet and pythonSet1:

```
>>> pythonSet.union(pythonSet1)
```

Here is the output:

```
set(['Pencil', 'Pen', 'NoteBook', 'Book', 'Eraser',  'Diary', 'Marker'])
```

Step 3-6-4. Performing an Intersection Operation on Sets

The intersection() function will return a new set with common elements from two given sets. We have already created two sets, pythonSet and pythonSet1. We can observe that Pencil and NoteBook are common elements in our sets. In following line, we use intersection() on our sets:

```
>>> pythonSet.intersection(pythonSet1)
```

Here is the output:

```
set(['Pencil', 'NoteBook'])
```

After running the code, it is clear that the intersection will return the elements that are common to both sets.

Recipe 3-7. Work with a Python Dictionary
Problem

You want to work with a Python dictionary.

Solution

You have seen that lists and tuples are indexed by their index numbers. This situation becomes clearer when the index uses words. A Python dictionary is a data structure that stores key/value pairs. Each element of a Python dictionary is a key/value pair. In this exercise, you want to do the following operations on a Python dictionary:

- Create a dictionary of stationary items, with the item ID as the key and the item name as the value
- Index an element using a key
- Get all the keys
- Get all the values

The creation of a dictionary can be achieved by using set of curly brackets, { }. You might be a little confused that we created a Python set using curly brackets and now we are going to create a dictionary in the same way. But let me tell you that, in order to create a dictionary, we have to provide a key/value pair inside the curly brackets—for example, {key:value}. We can observe that, in a key/value pair, the key is separated from the value by a colon (:). Two different key/value pairs are separated by a comma (,).

You can index a dictionary by using square brackets, [ ]. A Python dictionary object has a get() function, which returns the value for a given key.

You can get all keys by using the keys() function; the values() function returns all the values.

How It Works

We'll use the steps in this section to solve our problem.

Step 3-7-1. Creating a Dictionary of Stationary Items

We'll create a dictionary with the following line of Python code:

```
>>> pythonDict = {'item1':'Pencil','item2':'Pen', 'item3':'NoteBook'}
>>> pythonDict
```

Here is the output:

```
{'item2': 'Pen', 'item3': 'NoteBook', 'item1': 'Pencil'}
```

Step 3-7-2. Indexing an Element by Using a Key

Let's fetch the value of 'item1:'

```
>>> pythonDict['item1']
```

Here is the output:

```
'Pencil'
```

But if the key is not found in the dictionary, a KeyError exception is thrown:

```
>>> pythonDict['item4']
```

Here is the output:

```
Traceback (most recent call last):
  File "<stdin>", line 1, in <module>
KeyError: 'item4'
```

To prevent a KeyError exception, we can use the get() function on a dictionary. Then, if a value is not found for a given key, it returns nothing (and otherwise returns the value associated with that key). Let's explore get() in the following Python code snippet:

```
>>> pythonDict.get('item1')
```

Here is the output:

```
'Pencil'
```

We know that the key item4 is not in our dictionary. Therefore, the Python get() function will return nothing in this case:

```
>>> pythonDict.get('item4')
```

Step 3-7-3. Getting All the Keys

Getting all the keys together is often required in problem-solving. Let's get all the keys by using our keys() function:

```
>>> pythonDict.keys()
```

Here is the output:

```
['item2', 'item3', 'item1']
Function keys() returns all the keys in a list.
```

Step 3-7-4. Getting All the Values

Applying the value() function will return all the values in a dictionary:

```
>>> pythonDict.values()
```

Here is the output:

```
['Pen', 'NoteBook', 'Pencil']
```

Similar to the keys() function, values() also returns all the values in a list as output.

Recipe 3-8. Work with Define and Call Functions
Problem

You want to write a Python function that takes an integer as input and returns True if the input is an even number, and otherwise returns False.

Solution

Functions improve code readability and reusability. In Python, a function definition is started with the def keyword. The user-defined name of a function follows that def keyword. If a function takes an argument, it is written within parentheses; otherwise, we write just the set of parentheses. The colon symbol (:) follows the parentheses. Then we start writing the statements as the body of the function. The body is indented with respect to the line containing the def keyword. If our user-defined function returns a value, that is achieved by using the return Python keyword.

How It Works

We are going to create a function named isEvenInteger. The following lines define our function:

```
>>> def isEvenInteger(ourNum) :
        return ourNum %2 == 0
```

In this code example, we can see that our function name is isEvenInteger. Our function name follows the def keyword. The argument to our function, which is ourNum in this case, is within parentheses. After the colon, we define our function body. In this function body, there is only one statement. The function body statement is returning the output of a logical expression. We return a value from a function by using the return keyword. If a function doesn't return any value, the return keyword is omitted. The logical expression of the function body is checking whether modules of the input number equal zero or not.

After defining our function isEvenInteger, we should call it so we can watch it work. In order to test our function working, we'll call it with an even argument and then with an odd integer argument. Let's start with the even integer input. Here we provide input 4, which is an even integer:

```
>>> isEvenInteger(4)
```

Here is the output:

```
True
```

As we can see, the output is True. This means that the input we have provided is an even number.

Now let's provide an odd number as input:

```
>>> isEvenInteger(5)
```

Here is the output:

```
False
```

With an odd number as the input, our function isEvenInteger has returned False.

Recipe 3-9. Work with Create and Call Lambda Functions

Problem

You want to create a lambda function, which takes an integer as input and returns `True` if the input is an even number, and otherwise return `False`.

Solution

A *lambda function* is also known as an *anonymous function*. A lambda function is a function without a name. A lambda function in Python is defined with the `lambda` keyword.

How It Works

Let's create a lambda function that will check whether a given number is an even number:

```
>>> isEvenInteger = lambda ourNum : ourNum%2 == 0
```

Our lambda function is `lambda ourNum : ourNum%2 == 0`. You can see that we have used the keyword `lambda` to define it. After the keyword `lambda`, the arguments of the function have been written. Then after the colon, we have a logical expression. Our logical expression will check whether the input number is divisible by 2. In Python, functions are objects, so our lambda function is also an object. Therefore, we can put our object in a variable. That is how we can put our lambda function in the variable `isEvenInteger`:

```
>>> isEvenInteger(4
```

Here is the output:

```
True
```

```
>>> isEvenInteger(5)
```

Here is the output:

```
False
```

The drawback of lambda functions is that only one expression can be written. The result of that particular expression will be returned. The Python keyword `return` is not required to return the value of an expression. A lambda function cannot be extended beyond more than one line

Recipe 3-10. Work with Python Conditionals

Problem

You want to work with Python conditionals.

Solution

A car manufacturing company manufactures three types of cars. These three types of cars are differentiated by their number of cylinders. It is obvious that the different number of cylinders results in different gas mileage. Car A has four cylinders, car B has six cylinders, and car C has eight cylinders. Table 3-1 shows the number of cylinders and respective gas mileage of the cars.

Table 3-1. *Number of Cylinders and Respective Gas Mileage*

Number of Cylinders	Gas Mileage (Miles per Gallon)
4	22
6	18
8	16

Mr A, a sales executive, always forgets about this relationship between the number of cylinders and the gas mileage. So you want to create a small Python application that will return the mileage, given the number of cylinders. This will be a great help to Mr. A.

Conditionals are imperative in solving any programming problem. Python is no exception. Conditionals are implemented in Python by using if, elif, and else keywords.

How It Works

To create the required Python application, we are going to write a Python function that will take the number of cylinders as input and return the associated mileage.

Here's the function we'll use to create our application:

```
>>> def mpgFind(numOfCylinders) :
...       if(numOfCylinders == 4 ):
...            mpg = 22
...       elif(numOfCylinders == 6 ):
...            mpg = 18
...       else :
...            mpg = 16
...       return mpg
```

Here we have defined a function named mpgFind. Our function will take the variable numOfCylinders. Entering into function our variable going to face a logical expression numOfCylinders == 4. If numOfCylinders has the value 4, the logical expression will return True and the if block will be executed. Our if block has only one statement, mpg=22. If required, more than one statement can be provided.

If numOfCylinders == 4 results in False, the elif logical expression will be tested. If the value of numOfCylinders is 6, the logical expression of elif will return True, and 18 will be assigned to mpg.

What if the logical expression of if and elif both return False? In this condition, the else block will be executed. The variable mpg will be assigned 16.

Let's call our function with the input 4 and see the result:

```
>>> mpgFind(4)
```

Here is the output:

```
22
```

You can clearly see that our function can help our sales executive, Mr A.

Recipe 3-11. Work with Python "for" and "while" Loops
Problem
You want to work with the Python for and while loops.

Solution

After learning that his application was written in Python, sales executive Mr A became interested in learning Python. He joined a Python class. His instructor, Mr X, gave him an assignment to solve. Mr X asked his class to implement a Python function that will take a list of integers from 1 to 101 as input and then return the sum of the even numbers in the list. Mr A found that the required function can be implemented using the following:

- for loop
- while loop

Loops are best for code reusability. We use a loop to run a segment of code many times. A particular segment of code can be run using for and while loops. To get a summation of even numbers in a list, we require two steps: first we have to find the even numbers, and then we have to do the summations.

How It Works
Step 3-11-1. Implementing a for Loop

First, we are going to get the summation of even numbers in a list by using a for loop in Python:

```
>>> def sumUsingForLoop(numbers) :
...        sumOfEvenNumbers = 0
...        for i in numbers :
...            if i % 2 ==0 :
...                sumOfEvenNumbers = sumOfEvenNumbers +i
...            else :
...                pass
...        return    sumOfEvenNumbers
```

Here we have a written a Python function named sumUsingForLoop. Our function takes as input numbers. A for loop in Python iterates over a Python sequence. If we send a list of integers, the for loop starts iterating element by element. The if block will check whether the number being considered is an even number. If the number is even, the variable sumOfEvenNumbers will be increased by this number, using summation. After completion of the iteration, the final sum will be returned.

Let's check the working of our function:

```
>>> numbers = range(1,102)
>>> numbers
```

Here is the output:

```
[1, 2, 3, 4, 5, 6, 7,............,  98, 99, 100, 101]
```

We have created a list of integers, from 1 to 101, using the range() function. Let's call our function sumUsingForLoop and see the result:

```
>>> sumUsingForLoop(numbers)
```

Here is the output:

```
2550
```

Step 3-11-2. Implementing a while Loop

In this section, we are going to implement a solution to our problem that uses a while loop:

```
>>> def sumUsingWhileLoop(numbers) :
...        i = 1
...        sumOfEvenNumbers = 0
...        while i <= 101 :
```

```
...             if i % 2 ==0 :
...                 sumOfEvenNumbers = sumOfEvenNumbers +i
...             i = i + 1
...         return  sumOfEvenNumbers
```

Here we have defined a Python function named sumUsingWhileLoop. Our function takes as input numbers. A while loop in Python helps to iterate over a Python sequence. If we send a list of integers, we can iterate over our list, element by element, as shown in the code. The if block will check whether the number in consideration is an even number. If the number is even, the variable sumOfEvenNumbers will be increased by this number, using summation. After completion of the iteration, the final sum will be returned.

Let's test our function:

```
>>> sumUsingWhileLoop(numbers)
```

Here is the output:

```
2550
```

Recipe 3-12. Work with NumPy
Problem

You want to work with NumPy.

Solution

Company XYZ wants to build its new factory at location A. The company needs a location that has a temperature meeting certain specific criteria. Environmental scientist Mr. Y gathers the temperature reading at site A for five days at different times.

Table 3-2 depicts the temperature readings.

Table 3-2. *Temperatures in Celsius*

	6am	8am	10am	12am	2pm	4pm	6pm	8pm
day1	15	16	17	17	18	17	16	14
day2	14	15	17	17	16	17	16	15
day3	16	15	17	18	17	16	15	14
day4	16	17	18	19	17	15	15	14
day5	16	15	17	17	17	16	15	13

You want to do the following:

- Install pip

- Install NumPy

- Create a two-dimensional array using the NumPy array() function

- Create a two-dimensional array using vertical and column stacking of smaller arrays

- Know and change the data type of array elements

- Know shape of a given array

- Calculate minimum and maximum temperature for each day

- Calculate minimum and maximum temperature column-wise

- Calculate the mean temperature for each day and column-wise

- Calculate the standard deviation of temperature for each day and column-wise

- Calculate the variance of temperature for each day and column-wise

- Calculate the median temperature for each day and column-wise

- Calculate the overall mean from all the gathered temperature data

- Calculate the variance and standard deviation over all five days of temperature data

These sorts of simple mathematical questions can be solved easily by using NumPy. You might be thinking that we can solve these problems by using nested lists, so why are we going to use NumPy? Looping becomes faster with the NumPy ndarray. NumPy is open source and easily available.

The NumPy ndarray is a higher-level abstraction for multidimensional array data. It also provides many functions for working on those multidimensional arrays. In order to work with NumPy, we first have to install it. After creating a two-dimensional array of given temperature data, we can apply many NumPy functions to solve the given problems.

How It Works

We'll use the following steps to solve our problem.

Step 3-12-1. Installing pip

pip is a Python package management system that is written in Python itself. We can use pip to install other Python packages. Using the yum installer, we can install pip as follows:

```
[pysparkbook@localhost ~]$ sudo yum install python-pip
```

After installing pip, we have to install pyparsing. Run the following command to install pyparsing:

```
[pysparkbook@localhost ~]$ sudo  yum install ftp://mirror.switch.ch/pool/4/
mirror/centos/7.3.1611/cloud/x86_64/openstack-kilo/common/pyparsing-2.0.3-1.
el7.noarch.rpm
```

Step 3-12-2. Installing NumPy

After installing pip, it can be used to install NumPy. The following command installs NumPy on our machine:

```
[pysparkbook@localhost ~]$ sudo pip install numpy
```

Here is the output:

```
Collecting numpy
  Downloading numpy-1.12.0-cp27-cp27mu-manylinux1_x86_64.whl (16.5MB)
    100% |████████████████████████████████| 16.5MB 64kB/s
Installing collected packages: numpy
Successfully installed numpy-1.12.0
```

Step 3-12-3. Creating a Two-Dimensional Array by Using array( )

A multidimensional array can be created in many ways. In this step, we are going to create a two-dimensional array by using a nested Python list. So let's start creating a daily list of temperature data. In the following chunk of code, we are creating five lists for five days of temperatures:

```
>>> import numpy as NumPy
>>> temp1 = [15, 16, 17, 17, 18, 17, 16, 14]
>>> temp2 = [14, 15, 17, 17, 16, 17, 16, 15]
>>> temp3 = [16, 15, 17, 18, 17, 16, 15, 14]
>>> temp4 = [16, 17, 18, 19, 17, 15, 15, 14]
>>> temp5 = [16, 15, 17, 17, 17, 16, 15, 13]
```

The variable temp1 has the temperature measurements of the first day. Similarly, temp2, temp3, temp4, and temp5 have measurements of the temperature on the second, third, fourth, and fifth day, respectively.

Our two-dimensional array of temperatures can be created by using the NumPy array( ) function as follows:

```
>>> dayWiseTemp = NumPy.array([temp1,temp2,temp3,temp4,temp5])
>>> dayWiseTemp
```

Here is the output:

```
array([[15, 16, 17, 17, 18, 17, 16, 14],
       [14, 15, 17, 17, 16, 17, 16, 15],
       [16, 15, 17, 18, 17, 16, 15, 14],
       [16, 17, 18, 19, 17, 15, 15, 14],
       [16, 15, 17, 17, 17, 16, 15, 13]])
```

Now we have created a two-dimensional array.

Step 3-12-4. Creating a Two-Dimensional Array by Stacking

We can create an array by using vertical stacking and column stacking of data. First, we are going to create our same temperature array data by using vertical stacking. Vertical stacking can be created by using the NumPy vstack() function:

```
>>> dayWiseTemp = NumPy.vstack((temp1,temp2,temp3,temp4,temp5))
>>> dayWiseTemp
```

Here is the output:

```
array([[15, 16, 17, 17, 18, 17, 16, 14],
       [14, 15, 17, 17, 16, 17, 16, 15],
       [16, 15, 17, 18, 17, 16, 15, 14],
       [16, 17, 18, 19, 17, 15, 15, 14],
       [16, 15, 17, 17, 17, 16, 15, 13]])
```

Now let's see how to do horizontal stacking. Temperature data has been collected at different times and on different days. Let's now create a temperature data list based on time:

```
>>> d6am = NumPy.array([15, 14, 16, 16, 16])
>>> d8am = NumPy.array([16, 15, 15, 17, 15])
>>> d10am = NumPy.array([17, 17, 17, 18, 17])
>>> d12am = NumPy.array([17, 17, 18, 19, 17])
>>> d2pm = NumPy.array([18, 16, 17, 17, 17])
>>> d4pm = NumPy.array([17, 17, 16, 15, 16])
>>> d6pm = NumPy.array([16, 16, 15, 15, 15])
>>> d8pm = NumPy.array([14, 15, 14, 14, 13])
```

Column stacking can be done with the NumPy column_stack() function as follows:

```
>>> dayWiseTemp = NumPy.column_stack((d6am,d8am,d10am,d12am,d2pm,d4pm,d6pm,d8pm))
>>> dayWiseTemp
```

Here is the output:

```
array([[15, 16, 17, 17, 18, 17, 16, 14],
       [14, 15, 17, 17, 16, 17, 16, 15],
       [16, 15, 17, 18, 17, 16, 15, 14],
       [16, 17, 18, 19, 17, 15, 15, 14],
       [16, 15, 17, 17, 17, 16, 15, 13]])
```

Step 3-12-5. Knowing and Changing the Data Type of Array Elements

The NumPy array dtype attribute will return the data type of a NumPy array:

```
>>> dayWiseTemp.dtype
```

Here is the output:

```
dtype('int64')
```

We can observe that NumPy has inferred the data type as int64. From the previous historical temperature data of the given location, we know that the temperature will generally vary between 0 and 100. Therefore, using a 64-bit integer is not efficient. We can use 32-bit integers, which will take less memory than 64-bit integers.

In order to create an array with the data type int32, we can provide the dtype argument for the array() function:

```
>>> dayWiseTemp = NumPy.array([temp1,temp2,temp3,temp4,temp5],dtype='int32')
>>> dayWiseTemp.dtype
```

Here is the output:

```
dtype('int32')
```

But if the array has been created using the default data type, no worries. The data type of the array elements can be changed by using the NumPy astype() function. We can change the data type of an existing array by using the astype function as follows:

```
>>> dayWiseTemp = dayWiseTemp.astype('int32')
>>> dayWiseTemp.dtype
```

Here is the output:

```
dtype('int32')
```

Step 3-12-6. Knowing the Dimensions of an Array

The shape of an array can be calculated by using the shape attribute of the array:

```
>>> dayWiseTemp.shape
```

Here is the output:

```
(5, 8)
```

The output clearly indicates that our array has five rows and eight columns.

Step 3-12-7. Calculating the Minimum and Maximum Temperature Each Day

For a NumPy array, we can use the min() function to calculate the minimum. We can calculate the minimum of an array's data either by row or by column. To calculate the minimum temperature value of a row, we have to set the value of the axis argument to 1.

In our case, the data in a row indicates the temperature during one day. The following line of code will compute the minimum temperature value during a day:

```
>>> dayWiseTemp.min(axis=1)
```

Here is the output:

```
array([14, 14, 14, 14, 13], dtype=int32)
```

In similar fashion, the daily maximum temperature can be calculated by using the NumPy array max() function:

```
>>> dayWiseTemp.max(axis=1)
```

Here is the output:

```
array([18, 17, 18, 19, 17], dtype=int32)
```

Step 3-12-8. Calculating the Minimum and Maximum Temperature by Column

We can get the minimum and maximum temperature of a column by using the same min() and max() functions, respectively. But now we have to set the axis argument to 0, as follows:

```
>>> dayWiseTemp.min(axis=0)
```

Here is the output:

```
array([14, 15, 17, 17, 16, 15, 15, 13], dtype=int32)
>>> dayWiseTemp.max(axis=0)
```

Here is the output:

```
array([16, 17, 18, 19, 18, 17, 16, 15], dtype=int32)
```

Step 3-12-9. Calculating the Mean Temperature Each Day and by Column

Now we are going to calculate the mean value of the daily temperatures. Let's start by calculating the mean temperature of a day:

```
>>> dayWiseTemp.mean(axis=1)
```

Here is the output:

```
array([ 16.25,  15.875,  16.,  16.375,  15.75 ])
```

In order to calculate the mean temperature by column, we have to set the axis argument to 0:

```
>>> dayWiseTemp.mean(axis=0)
```

Here is the output:

```
array([ 15.4,  15.6,  17.2,  17.6,  17.,  16.2,  15.4,  14. ])
```

Step 3-12-10. Calculating the Standard Deviation of Temperature Each Day

Let's start with calculating the daily standard deviation of temperature. We are going to calculate the standard deviation by using the std() function of the ndarray class:

```
>>> dayWiseTemp.std(axis=1)
```

Here is the output:

```
array([ 1.19895788,  1.05326872,  1.22474487,  1.57619003,  1.29903811])
```

The column's standard deviation can be calculated by using the std() function with the value of axis set to 0:

```
>>> dayWiseTemp.std(axis=0)
```

Here is the output:

```
array([ 0.8,   0.8,   0.4,   0.8,   0.63245553,
        0.74833148,   0.48989795,   0.63245553])
```

Step 3-12-11. Calculating the Variance of Temperature Each Day

The NumPy array var() function can be used to calculate variance per row. Let's start with calculating the daily temperature variance:

```
>>> dayWiseTemp.var(axis=1)
```

Here is the output:

```
array([ 1.4375,  1.109375,  1.5,  2.484375,  1.6875  ])
```

The temperature variance of columns can be calculated as follows:

```
>>> dayWiseTemp.var(axis=0)
```

Here is the output:

```
array([ 0.64,   0.64,   0.16,   0.64,   0.4 ,   0.56,   0.24,   0.4 ])
```

Step 3-12-12. Calculating Daily and Hourly Medians

The median can be calculated by using the NumPy median() function as follows:

```
>>>NumPy.median(dayWiseTemp,axis=1)
```

Here is the output:

```
array([ 16.5,   16. ,   16. ,   16.5,   16. ])
>>> NumPy.median(dayWiseTemp,axis=0)
```

Here is the output:

```
array([ 16.,   15.,   17.,   17.,   17.,   16.,   15.,   14.])
```

Step 3-12-13. Calculating the Overall Mean of all the Gathered Temperature Data

The NumPy mean() function can be used to calculate the overall mean of all data:

```
>>> NumPy.mean(dayWiseTemp)
```

Here is the output:

```
16.050000000000001
```

Step 3-12-14. Calculating the Variance and Standard Deviation over All Five Days of Temperature Data

The NumPy var() function can be used to calculate the variance of all the gathered data:

```
>>> NumPy.var(dayWiseTemp)
```

Here is the output:

```
1.6974999999999993
```

And the std() function can be used to calculate the standard deviation:

```
>>> NumPy.std(dayWiseTemp)
```

Here is the output:

```
1.3028814220795379
```

▪ **Note**   You can learn more about NumPy at www.numpy.org.

Recipe 3-13. Integrate IPython and IPython Notebook with PySpark
Problem

You want to integrate IPython and IPython Notebook with PySpark.

Solution

Integrating IPython with PySpark improves programmer efficiency. To accomplish this integration, we'll do the following:

- Install IPython
- Integrate PySpark with IPython
- Install IPython Notebook
- Integrate PySpark with IPython Notebook
- Run PySpark commands on IPython Notebook

You have become familiar with the Python interactive shell. This interactive shell enables us to learn Python faster, because we can see the result of each command, line by line. The Python shell that comes with Python is very basic. It does not come with tab completion and other features. A more advanced Python interactive shell is IPython. It has many advanced features that facilitate coding.

IPython Notebook start a web-browser-based facility to write Python code. Generally, readers may be confused that for a web-browser-based notebook, we need an Internet connection. That's not the case; we can run IPython Notebook without an Internet connection.

We can start PySpark with IPython and IPython Notebook.

It is time to install IPython and IPython Notebook and integrate PySpark with those. IPython and IPython Notebook can be installed using pip.

How It Works

In order to connect PySpark, we have to perform the following steps.

Step 3-13-1. Installing IPython

Let's install IPython first. We have already installed pip. Pip can be used to install IPython with the following command:

```
[pysparkbook@localhost ~]$ sudo pip install ipython
```

Here is the output:

```
Collecting ipython
Successfully installed ipython-5.2.2 pathlib2-2.2.1 pexpect-4.2.1
pickleshare-0.7.4 prompt-toolkit-1.0.13 ptyprocess-0.5.1 scandir-1.4
simplegeneric-0.8.1 wcwidth-0.1.7
```

Step 3-13-2. Integrating PySpark with IPython

After installation, we are going to start PySpark with IPython. This is easy to do. First, we set the environmental variable IPYTHON equal to 1, as follows:

```
[pysparkbook@localhost ~]$ export IPYTHON=1
[pysparkbook@localhost ~]$ pyspark
```

After starting PySpark, you'll see the shell in Figure 3-1.

Figure 3-1. *Shell*

In Figure 3-1, you can see that now In[1] has replaced the legacy >>> symbol of our old Python console.

Step 3-13-3. Installing IPython Notebook

Now we are going to install IPython Notebook. Again, let's use pip to install IPython:

```
[pysparkbook@localhost ~]$ sudo pip install ipython[notebook]
```

Step 3-13-4. Integrating PySpark with IPython Notebook

After installation, we have to set some environment variables:

```
[pysparkbook@localhost ~]$ export IPYTHON_OPTS="notebook"
[pysparkbook@localhost ~]$ export XDG_RUNTIME_DIR=""
```

Now it is time to start PySpark:

```
[pysparkbook@localhost ~]$ pyspark
```

After starting PySpark with IPython Notebook, we'll see the web browser shown in Figure 3-2.

Figure 3-2. *After starting PySpark with Notebook, you'll see a web browser*

You might be amazed that the browser is showing *Jupyter*. But this isn't amazing; Jupyter is the new name for IPython Notebook. You can see how easy it is to work with Notebook; by clicking the Upload button, we can upload files to the machine.

Step 3-13-5. Running PySpark Commands on IPython Notebook

In order to run the PySpark command, let's create a new notebook by using Python 2, as shown in Figure 3-3.

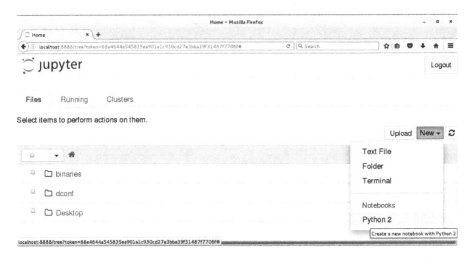

Figure 3-3. *Creating a new notebook using Python 2*

After creating the notebook, you will see the web page in Figure 3-4.

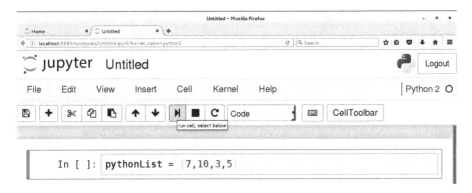

Figure 3-4. *After creating the notebook, you'll see this web page*

Now we can run the Python commands to create a list. After writing our command, we have to run it. You can see in Figure 3-4 that we can run our command by using the Run button in the notebook.

In Figure 3-5, we are printing pythonList.

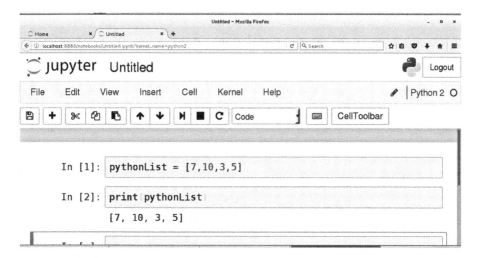

Figure 3-5. *Printing the Python list*

CHAPTER 4

■ ■ ■

Spark Architecture and the Resilient Distributed Dataset

You learned Python in the preceding chapter. Now it is time to learn PySpark and utilize the power of a distributed system to solve problems related to big data. We generally distribute large amounts of data on a cluster and perform processing on that distributed data.

This chapter covers the following recipes:

> Recipe 4-1. Create an RDD

> Recipe 4-2. Convert temperature data

> Recipe 4-3. Perform basic data manipulation

> Recipe 4-4. Run set operations

> Recipe 4-5. Calculate summary statistics

> Recipe 4-6. Start PySpark shell on Standalone cluster manager

> Recipe 4-7. Start PySpark shell on Mesos

> Recipe 4-8. Start PySpark shell on YARN

Learning about the architecture of Spark will be very helpful to your understanding of the various components of Spark. Before delving into the recipes let's explore this topic.

Figure 4-1 describes the Spark architecture.

© Raju Kumar Mishra 2018
R. K. Mishra, *PySpark Recipes*, https://doi.org/10.1007/978-1-4842-3141-8_4

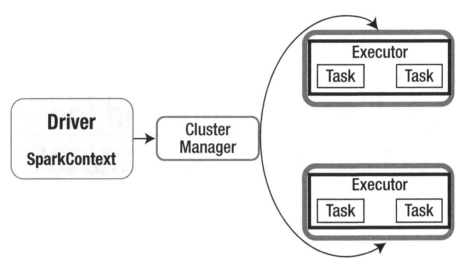

Figure 4-1. Spark architecture

The main components of the Spark architecture are the *driver* and *executors*. For each PySpark application, there will be one driver program and one or more executors running on the cluster slave machine. You might be wondering, what is an application in the context of PySpark? An *application* is a whole bunch of code used to solve a problem.

The *driver* is the process that coordinates with many executors running on various slave machines. Spark follows a master/slave architecture. The SparkContext object is created by the driver. SparkContext is the main entry point to a PySpark application. You will learn more about SparkContext in upcoming chapters. In this chapter, we will run our PySpark commands in the PySpark shell. After starting the shell, we will find that the SparkContext object is automatically created. We will encounter the SparkContext object in the PySpark shell as the variable sc. The shell itself is working as our driver. The driver breaks our application into small tasks; a *task* is the smallest unit of your application. Tasks are run on different executors in parallel. The driver is also responsible for scheduling tasks to different executors.

Executors are slave processes. An executor runs tasks. It also has the capability to cache data in memory by using the BlockManager process. Each executor runs in its own Java Virtual Machine (JVM).

The cluster manager manages cluster resources. The driver talks to the cluster manager to negotiate resources. The cluster manager also schedules tasks on behalf of the driver on various slave executor processes. PySpark is dispatched with Standalone Cluster Manager. PySpark can also be configured on YARN and Apache Mesos. In our recipes, you are going to see how to configure PySpark on Standalone Cluster Manager and Apache Mesos. On a single machine, PySpark can be started in local mode too.

The main celebrated component of PySpark is the *resilient distributed dataset* (RDD). The RDD is a data abstraction over the distributed collection. Python collections such as lists, tuples, and sets can be distributed very easily. An RDD is recomputed on node failures. Only part of the data is calculated or recalculated, as required. An RDD is created using various functions defined in the SparkContext class. One important method for creating an RDD is `parallelize()`, which you will encounter again and again in this chapter. Figure 4-2 illustrates the creation of an RDD.

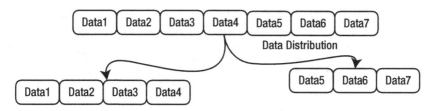

Figure 4-2. *Creating an RDD*

Let's say that we have a Python collection with the elements Data1, Data2, Data3, Data4, Data5, Data6, and Data7. This collection is distributed over the cluster to create an RDD. For simplicity, we can assume that two executors are running. Our collection is divided into two parts. The first executor gets the first part of the collection, which has the elements Data1, Data2, Data3, and Data4. The second part of the collection is sent to the second executor. So, the second executor has the data elements Data5, Data6, and Data7.

We can perform two types of operations on the RDD: transformation and action. *Transformation* on an RDD returns another RDD. We know that RDDs are immutable; therefore, changing the RDD is impossible. Hence transformations always return another RDD. Transformations are lazy, whereas actions are eagerly evaluated. I say that the transformation is *lazy* because whenever a transformation is applied to an RDD, that operation is not applied to the data at the same time. Instead, PySpark notes the operation request, but all the transformations are applied when the first action is called.

Figure 4-3 illustrates a transformation operation. The transformation on RDD1 creates RDD2. RDD1 has two partitions. The first partition of RDD1 has four data elements: Data1, Data2, Data3, and Data4. The second data partition of RDD1 has three elements: Data5, Data6, and Data7. After transformation on RDD1, RDD2 is created. RDD2 has six elements. So it is clear that the daughter RDD might have a different number of data elements than the father RDD. RDD2 also has two partitions. The first partition of RDD2 has three data points: Data8, Data9, and Data10. The second partition of RDD2 also has three elements: Data11, Data12, and Data13. Don't get confused about the daughter RDD having a different number of partitions than the father RDD.

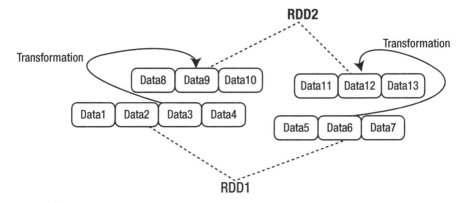

Figure 4-3. *RDD transformations*

Figure 4-4 illustrates an action performed on an RDD. In this example, we are applying the summation action. Summed data is returned to the driver. In other cases, the result of an action can be saved to a file or to another destination.

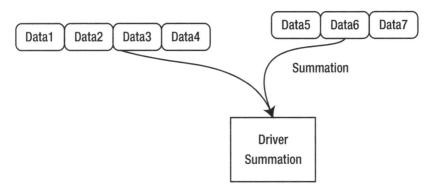

Figure 4-4. *RDD action*

You might be wondering, if Spark has been written in Scala, then how is Python contacting with Scala? You might guess that a Python wrapper of PySpark has been written using Jython, and that this Jython code is compiled to Java bytecode and run on the JVM. This guess isn't correct.

A running Python program can access Java objects in a JVM by using Py4J. A running Java program can also access Python objects by using Py4J. A gateway between Python and Java enables Python to use Java objects.

Driver programs use Py4J to communicate between Python and the Java SparkContext object. PySpark uses Py4J, so that PySpark Python code can

On remote cluster machines, the PythonRDD object creates Python subprocesses and communicates with them using pipes. The PythonRDD object runs in JVM and communicates with Python processes by using pipes.

■ **Note**   You can learn more about Py4J at the following locations:

www.py4j.org

https://cwiki.apache.org/confluence/display/SPARK/PySpark+Internals

---

Recipe 4-1. Create an RDD

Problem

You want to create an RDD.

Solution

As we know, an RDD is a distributed collection. You have a list with the following data:

pythonList = [2.3,3.4,4.3,2.4,2.3,4.0]

You want to do the following operations:

- Create an RDD of the list
- Get the first element
- Get the first two elements
- Get the number of partitions in the RDD

In PySpark, an RDD can be created in many ways. One way to create an RDD out of a given collection is to use the parallelize() function. The SparkContext object is used to call the parallelize() function. You'll read more about SparkContext in an upcoming chapter.

In the case of big data, even tabular data, a table might have more than 1,000 columns. Sometimes analysts want to see what those columns of data look like. The first() function is defined on an RDD and will return the first element of the RDD.

To get more than one element from a list, we can use the take() function. The number of partitions of a collection can be fetched by using getNumPartitions().

How It Works

Let's follow the steps in this section to solve the problem.

Step 4-1-1. Creating an RDD of the List

Let's first create a Python list by using the following:

```
>>> pythonList = [2.3,3.4,4.3,2.4,2.3,4.0]
>>> pythonList
```

Here is the output:

```
[2.3, 3.4, 4.3, 2.4, 2.3, 4.0]
```

Parallelization or distribution of data is done using the `parallelize()` function. This function takes two arguments. The first argument is the collection to be parallelized, and the second argument indicates the number of distributed chunks of data you want:

```
>>> parPythonData = sc.parallelize(pythonList,2)
```

Using the `parallelize()` function, we have distributed our data in two partitions. In order to get all the data on the driver, we can use the `collect()` function, as shown in the following code line. Using the `collect()` function is not recommended in production; rather, it should be used only in code debugging.

```
>>> parPythonData.collect()
```

Here is the output:

```
[2.3, 3.4, 4.3, 2.4, 2.3, 4.0]
```

Step 4-1-2. Getting the First Element

The `first()` function can be used to get the first data out of an RDD. You might have figured out that the `collect()` and `first()` functions perform actions:

```
>>> parPythonData.first()
```

Here is the output:

```
2.3
```

Step 4-1-3. Getting the First Two Elements

Sometimes data analysts want to see more than one row of data. The `take()` function can be used to fetch more than one row from an RDD. The number of rows you want is given as an argument to the `take()` function:

```
>>> parPythonData.take(2)
```

Here is the output:

```
[2.3, 3.4]
```

Step 4-1-4. Getting the Number of Partitions in the RDD

In order to optimize PySpark code, a proper distribution of data is required. The number of partitions of an RDD can be found using the getNumPartitions() function:

```
>>> parPythonData.getNumPartitions()
```

Here is the output:

```
2
```

Recall that we were partitioning our data into two partitions while using the parallelize() function.

Recipe 4-2. Convert Temperature Data
Problem

You want to convert temperature data by writing a temperature unit conversion program on an RDD.

Solution

You are given daily temperatures in Fahrenheit. You want to perform some analysis on that data. But your new software takes input in Celsius. Therefore, you want to convert your temperature data from Fahrenheit to Celsius. Table 4-1 shows the data you have.

Table 4-1. *Daily Temperature in Fahrenheit*

	day1	day2	day3	day4	day5	day6	day7
Temp In °F	59	57.2	53.6	55.4	51.8	53.6	55.4

You want to do the following:

- Convert temperature from Fahrenheit to Celsius
- Get all the temperature data points greater than 13° C

We can convert temperature from Fahrenheit to Celsius by using the following mathematical formula:

$$°C = (°F - 32) \times 5/9$$

We can see that in PySpark, this is a transformation problem. We can achieve this task by using the map() function on the RDD.

Getting all the temperatures greater than 13° C is a filtering problem. Filtering of data can be done by using the filter() function on the RDD.

How It Works

We'll follow the steps in this section to complete the conversion and filtering exercises.

Step 4-2-1. Parallelizing the Data

We are going to parallelize data by using our parallelize() function. We are going to distribute our data in two partitions, as follows:

```
>>> tempData = [59,57.2,53.6,55.4,51.8,53.6,55.4]
>>> parTempData = sc.parallelize(tempData,2)
>>> parTempData.collect()
```

Here is the output:

```
[59, 57.2, 53.6, 55.4, 51.8, 53.6, 55.4]
```

The collection of data has returned our parallelized data.

Step 4-2-2. Converting Temperature from Fahrenheit to Celsius

Now we are going to convert our temperature in Fahrenheit to Celsius. We'll write a fahrenheitToCentigrade function, which will take the temperature in Fahrenheit and return a temperature in Celsius for a given input:

```
>>> def fahrenheitToCentigrade(temperature) :
...    centigrade = (temperature-32)*5/9
...    return centigrade
```

Let's test our fahrenheitToCentigrade function:

```
>>> fahrenheitToCentigrade(59)
```

Here is the output:

15

We are providing 59 as the input in Fahrenheit. Our function returns a Celsius value of our Fahrenheit input; 59° F is equal to 15° C.

```
>>> parCentigradeData = parTempData.map(fahrenheitToCentigrade)
```

```
>>> parCentigradeData.collect()
```

Here is the output:

```
[15, 14.000000000000002, 12.0, 13.0, 10.999999999999998, 12.0, 13.0]
```

We have converted the given temperature to Celsius. Now let's filter out all the temperatures greater than or equal to 13° C.

Step 4-2-3. Filtering Temperatures Greater than 13° C

To filter data, we can use the filter() function on the RDD. We have to provide a predicate as input to the filter() function. A *predicate* is a function that tests a condition and returns True or False.

Let's define the predicate tempMoreThanThirteen, which will take a temperature value and return True if input is greater than or equal to 13:

```
>>> def tempMoreThanThirteen(temperature):
...   return temperature >=13
```

We are going to send our tempMoreThanThirteen function as input to the filter() function. The filter() function will iterate over each value in the parCentigradeData RDD. For each value, the tempMoreThanThirteen function will be applied. If the value is greater than or equal to 13, True will be returned. The value for which tempMoreThanThirteen returns True will come to filteredTemprature:

```
>>> filteredTemprature = parCentigradeData.filter(tempMoreThanThirteen)
```

```
>>> filteredTemprature.collect()
```

Here is the output:

```
[15, 14.000000000000002, 13.0, 13.0]
```

We can replace our predicates by using the lambda function. (We discussed lambda functions in Chapter 3.) Using a lambda function makes the code more readable. The following code line clearly depicts that the filter() function takes a predicate as input and returns True for all the values greater than or equal to 13:

```
>>> filteredTemprature = parCentigradeData.filter(lambda x : x>=13)
```

```
>>> filteredTemprature.collect()
```

Here is the output:

```
[15, 14.000000000000002, 13.0, 13.0]
```

We finally have four elements indicating a temperature that is either greater than or equal to 13. So now you understand the way to do basic analysis on data with PySpark.

Recipe 4-3. Perform Basic Data Manipulation
Problem

You want to do data manipulation and run aggregation operations.

Solution

In this recipe, you are given data indicating student grades for a two-year (four-semester) course. Seven students are enrolled in this course. Table 4-2 depicts two years of grade data, divided into semesters, for seven enrolled students.

Table 4-2. *Student Grades*

Student ID	Year	Semester1 Marks	Semester2 Marks
si1	year1	62.08	62.4
si1	year2	75.94	76.75
si2	year1	68.26	72.95
si2	year2	85.49	75.8
si3	year1	75.08	79.84
si3	year2	54.98	87.72
si4	year1	50.03	66.85
si4	year2	71.26	69.77
si5	year1	52.74	76.27
si5	year2	50.39	68.58
si6	year1	74.86	60.8
si6	year2	58.29	62.38
si7	year1	63.95	74.51
si7	year2	66.69	56.92

You want to calculate the following:

- Average grades per semester, each year, for each student

- Top three students who have the highest average grades in the second year

- Bottom three students who have the lowest average grades in the second year

- All students who have earned more than an 80% average in the second semester of the second year

Using the map() function is often helpful. In this example, the average grades per semester, for each year, can be calculated using map().

It is a general data science problem to get the top k elements, such as the top k highly performing bonds. The PySpark takeOrdered() function is going to take the top k or top bottom elements from our RDD.

Students who have earned more than 80% averages in the second year can be filtered using the filter() function.

How It Works

Let's solve our problem in steps. We will start with creating an RDD of our data.

Step 4-3-1. Making a List from a Given Table

In this step, we'll create a nested list. This means that each element of the list is a record, and each record is a list in itself:

```
>>> studentMarksData = [["si1","year1",62.08,62.4],
...    ["si1","year2",75.94,76.75],
...    ["si2","year1",68.26,72.95],
...    ["si2","year2",85.49,75.8],
...    ["si3","year1",75.08,79.84],
...    ["si3","year2",54.98,87.72],
...    ["si4","year1",50.03,66.85],
...    ["si4","year2",71.26,69.77],
...    ["si5","year1",52.74,76.27],
...    ["si5","year2",50.39,68.58],
...    ["si6","year1",74.86,60.8],
...    ["si6","year2",58.29,62.38],
...    ["si7","year1",63.95,74.51],
...    ["si7","year2",66.69,56.92]]
```

Step 4-3-2. Parallelizing the Data

After parallelizing the data by using the `parallelize()` function, we will find that we have an RDD in which each element is a list itself:

```
>>> studentMarksDataRDD = sc.parallelize(studentMarksData,4)
```

As we know, the `collect()` function takes the whole RDD to the driver. If the RDD size is very large, the driver may face a memory issue. In order to fetch k first elements of an RDD, we can use the `take()` function with n as input to `take()`. As an example, in the following line of code, we are fetching two elements of our RDD. Remember here that `take()` is an action:

```
>>> studentMarksDataRDD.take(2)
```

Here is the output:

```
[['si1', 'year1', 62.08, 62.4],
 ['si1', 'year2', 75.94, 76.75]]
```

Step 4-3-3. Calculating Average Semester Grades

Now let me explain what I want to do in the following code. Just consider the first element of the RDD. Our first element of the RDD is `['si1', 'year1', 62.08, 62.4]`, which is a list of four elements. Our work is to calculate the mean of grades from two semesters. In the first element, the mean is $0.5(62.08 + 62.4)$. We are going to use the `map()` function to get our solution.

```
>>> studentMarksMean = studentMarksDataRDD.map(lambda x :
[x[0],x[1],(x[2]+x[3])/2])
```

Again, we use the `take()` function to visualize the `map()` function output:

```
>>> studentMarksMean.take(2)
```

Here is the output:

```
[['si1', 'year1', 62.239999999999995],
 ['si1', 'year2', 76.345]]
```

Step 4-3-4. Filtering Student Average Grades in the Second Year

The following line of code is going to filter out all the data of the second year. We have implemented our predicate by using a lambda function. Our predicate function checks whether year2 is in the list. If the predicate returns `True`, the list includes second-year grades.

```
>>> secondYearMarks = studentMarksMean.filter(lambda x : "year2" in x)
```

```
>>> secondYearMarks.take(2)
```

Here is the output:

```
[['si1', 'year2', 76.345],
 ['si2', 'year2', 80.645]]
```

We can clearly see that the RDD output of secondYearMarks has only second-year grades.

Step 4-3-5. Finding the Top Three Students

We can get the top three students in two ways. The first method is to sort the full data according to grades. Obviously, we are going to sort the data in decreasing order. Sorting is done by the sortBy() function. Let's see the implementation:

```
>>> sortedMarksData = secondYearMarks.sortBy(keyfunc = lambda x : -x[2])
```

In our sortBy() function, we provide the keyfunc parameter. This parameter indicates to sort the grades data in decreasing order. Now collect the output and see the result:

```
>>> sortedMarksData.collect()
```

Here is the output:

```
[['si2', 'year2', 80.645],
 ['si1', 'year2', 76.345],
 ['si3', 'year2', 71.35],
 ['si4', 'year2', 70.515],
 ['si7', 'year2', 61.805],
 ['si6', 'year2', 60.335],
 ['si5', 'year2', 59.485]]
```

After sorting data, we can take the first three elements by using our take() function:

```
>>>  sortedMarksData.take(3)
```

Here is the output:

```
[['si2', 'year2', 80.645],
 ['si1', 'year2', 76.345],
 ['si3', 'year2', 71.35]]
```

We have our answer. But can we optimize it further? In order to get top-three data, we are sorting the whole list. We can optimize this by using the takeOrdered() function.

This function takes two arguments: the number of elements we require, and key, which uses a lambda function to determine how to take the data out.

```
>>> topThreeStudents = secondYearMarks.takeOrdered(num=3, key = lambda x :-x[2])
```

In the preceding code, we set num to 3 for the three top elements, and lambda in key so that it can provide three top in decreasing order.

```
>>> topThreeStudents
```

Here is the output:

```
[['si2', 'year2', 80.645],
 ['si1', 'year2', 76.345],
 ['si3', 'year2', 71.35]]
```

In order to print the result, we are not using the collect() function to get the data. Remember that transformation creates another RDD, so we require the collect() function to collect data. But an action will directly fetch the data to the driver, and collect() is not required. So you can conclude that the takeOrdered() function is an action.

Step 4-3-6. Finding the Bottom Three Students

We have to find the bottom three students in terms of their average grades. One way is to sort the data in increasing order and take the three on top. But that is not an efficient way, so we will use the takeOrdered() function again, but with a different key parameter:

```
>>> bottomThreeStudents = secondYearMarks.takeOrdered(num=3, key = lambda x :x[2]])
>>> bottomThreeStudents
```

Here is the output:

```
[['si5', 'year2', 59.485],
 ['si6', 'year2', 60.335],
 ['si7', 'year2', 61.805]]
```

Step 4-3-7. Getting All Students with 80% Averages

Now that you understand the filter() function, it is easy to guess that we can solve this problem by using filter(). We will have to provide a predicate, which will return True if grades are greater than 80; otherwise, it returns False.

```
>>> moreThan80Marks = secondYearMarks.filter(lambda x : x[2] > 80)
>>> moreThan80Marks.collect()
```

Here is the output:

```
[['si2', 'year2', 80.645]]
```

It can be observed that only one student (with the student ID si2) has secured more than an 80% average in the second year.

Recipe 4-4. Run Set Operations
Problem

You want to run set operations on a research company's data.

Solution

XYZ Research is a company that performs research on many diversified topics. Each research project comes with a research ID. Research may come to a conclusion in one year or may take more than one year. The following data is provided, indicating the number of research projects being conducted in three years:

```
2001: RIN1, RIN2, RIN3, RIN4, RIN5, RIN6, RIN7
2002: RIN3, RIN4, RIN7, RIN8, RIN9
2003: RIN4, RIN8, RIN10, RIN11, RIN12
```

Now we have to answer the following questions:

- How many research projects were initiated in the three years?

- How many projects were completed in the first year?

- How many projects were completed in the first two years?

A *set* is collection of distinct elements. PySpark performs pseudo set operations. They are called *pseudo set operations* because some functions do not remove duplicate elements.

Remember, the first question is not asking about completed projects. The total number of research projects initiated in three years is just the union of all three years of data. You can perform a union on two RDDs by using the union() function.

The projects that have been started in the first year and not in the second year are the projects that have been completed in the first year. Every project that is started is completed. We can use the subtract() function to find all the projects that were completed in the first year.

If we make a union of first-year and second-year projects and subtract third-year projects, we are going to get all the projects that have been completed in the first two years.

How It Works

Let's solve this problem step-by-step.

Step 4-4-1. Creating a List of Research Data by Year

Let's start with creating a list of all the projects that the company worked on each year:

```
>>> data2001 = ['RIN1', 'RIN2', 'RIN3', 'RIN4', 'RIN5', 'RIN6', 'RIN7']
>>> data2002 = ['RIN3', 'RIN4', 'RIN7', 'RIN8', 'RIN9']
>>> data2003 = ['RIN4', 'RIN8', 'RIN10', 'RIN11', 'RIN12']
```

data2001 is list of all the projects started in 2001. Similarly, data2002 contains all the research projects that either are continuing from 2001 or started in 2002. The data2003data list contains all the projects that the company worked on in 2003.

Step 4-4-2. Parallelizing the Data (Creating the RDD)

After creating lists, we have to parallelize our data:

```
>>> parData2001 = sc.parallelize(data2001,2)
>>> parData2002 = sc.parallelize(data2002,2)
>>> parData2003 = sc.parallelize(data2003,2)
```

After parallelizing, we get three RDDs. The first RDD is parData2001, the second RDD is parData2002, and the last one is parData2003.

Step 4-4-3. Finding Projects Initiated in Three Years

The total number of projects initiated in three years is determined just by getting the union of all the data for the given three years. RDD union() takes another RDD as input and returns, merging these two RDDs. Let's see how it works:

```
>>> unionOf20012002 = parData2001.union(parData2002)
>>> unionOf20012002.collect()
```

Here is the output:

```
['RIN1', 'RIN2', 'RIN3', 'RIN4',
 'RIN5', 'RIN6', 'RIN7', 'RIN3',
 'RIN4', 'RIN7', 'RIN8', 'RIN9']
```

We have calculated the union of different research projects initiated in either the first year or the second year. We can observe that the unionized data, unionOf20012002, has duplicate values. Having duplicates values in sets is not allowed. Therefore, a set operation on an RDD is also known as a pseudo set operation. Don't worry; we will remove these duplicates.

In order to get all the research projects that have been initiated in three years, we have to get the union of parData2003 and unionOf20012002:

```
>>> allResearchs = unionOf20012002.union(parData2003)

>>> allResearchs.collect()
```

Here is the output:

```
['RIN1', 'RIN2', 'RIN3', 'RIN4',
 'RIN5', 'RIN6', 'RIN7', 'RIN3',
 'RIN4', 'RIN7', 'RIN8', 'RIN9',
 'RIN4', 'RIN8', 'RIN10', 'RIN11', 'RIN12']
```

We have the union of all three years of data. Now we have to get rid of duplicates.

Step 4-4-4. Making Sets of Distinct Data

We are going to apply the distinct() function to our RDD allResearchs:

```
>>> allUniqueResearchs = allResearchs.distinct()
>>> allUniqueResearchs.collect()
```

Here is the output:

```
['RIN1', 'RIN12', 'RIN5', 'RIN3',
 'RIN4', 'RIN2', 'RIN11', 'RIN7',
 'RIN9', 'RIN6', 'RIN8', 'RIN10']
```

We can see that we have all the research projects that were initiated in the first three years.

Step 4-4-5. Counting Distinct Elements

Now count all the distinct research projects by using the count() function on the RDD:

```
>>> allResearchs.distinct().count()
```

Here is the output:

12

■ **Note**   We can run telescopic commands in PySpark too.

The following command is run in telescopic fashion:

```
>>> parData2001.union(parData2002).union(parData2003).distinct().count()
```

Here is the output:

```
12
```

Step 4-4-6. Finding Projects Completed the First Year

Let's say we have two sets, A and B. Subtracting set B from set A will give us all the elements that are members of set A but not set B. So now it is clear that, in order to know all the projects that have been completed in the first year (2001), we have to subtract the projects in year 2002 from all the projects in year 2001.

Subtraction on a set can be done with the subtract() function:

```
>>> firstYearCompletion = parData2001.subtract(parData2002)
>>> firstYearCompletion.collect()
```

Here is the output:

```
['RIN5', 'RIN1', 'RIN6', 'RIN2']
```

We have all the projects that were completed in 2001. Four projects were completed in 2001.

Step 4-4-7. Finding Projects Completed in the First Two Years

A union of RDDs gives us all the projects started in the first two years. After getting all the projects started in the first two years, if we then subtract projects running and started in the third year, we will return all the projects completed in the first two years. The following is the implementation:

```
>>> unionTwoYears = parData2001.union(parData2002)
>>> unionTwoYears.subtract(parData2003).collect()
```

Here is the output:

```
['RIN1', 'RIN5', 'RIN3', 'RIN3',
 'RIN2', 'RIN7', 'RIN7', 'RIN9', 'RIN6']
```

Now subtract:

```
>>> unionTwoYears.subtract(parData2003).distinct().collect()
```

Here is the output:

```
['RIN1', 'RIN5', 'RIN3',
 'RIN2', 'RIN7', 'RIN9', 'RIN6']
```

Step 4-4-8. Finding Projects Started in 2001 and Continued Through 2003.

This step requires using the intersection() method defined in PySpark on the RDD:

```
>>> projectsInTwoYear = parData2001.intersection(parData2002)
```

```
>>> projectsInTwoYear.collect()
```

Here is the output:

```
['RIN4', 'RIN7', 'RIN3']
```

```
>>> projectsInTwoYear.subtract(parData2003).distinct().collect()
```

Here is the output:

```
['RIN3', 'RIN7']
```

Recipe 4-5. Calculate Summary Statistics
Problem

You want to calculate summary statistics on given data.

Solution

Renewable energy sources are gaining in popularity all over the world. The company FindEnergy wants to install windmills at a given location. For efficient operation of windmills, the air requires certain characteristics.

Data is collected as shown in Table 4-3.

Table 4-3. *Air Velocity Data*

	8 am	10 am	12 am	2 pm	4 pm	6 pm	8 pm
Air Velocity in KMPH	12	13	15	12	11	12	11

You, as a data scientist, want to calculate the following quantities:

- Number of data points
- Summation of air velocities over a day
- Mean air velocity in a day
- Variance of air data
- Sample variance of air data
- Standard deviation of air data
- Sample standard deviation of air data

PySpark provides many functions to summarize data on the RDD. The number of elements in an RDD can be found by using the count() function on the RDD. There are two ways to sum all the data in a given RDD. The first is to apply the sum() method to the RDD. The second is to apply the reduce() function to the RDD.

The *mean* represents the center point of the given data, and it can be calculated in two ways too. We are going to use the mean() method and the fold() method to calculate the mean.

The *variance*, which indicates the spread of data around the mean, can be calculated using the variance() function. Similarly, the sample variance can be calculated by using the sampleVariance() method on the RDD.

Standard deviation and sample standard deviation will be calculated using the stdev() and sampleStdev() methods, respectively.

PySpark provides the stats() method, which can calculate all the previously mentioned quantities in one go.

How It Works

We'll follow the steps in this section to reach a solution.

Step 4-5-1. Parallelizing the Data

Let's parallelize the air velocity data from a list:

```
>>> airVelocityKMPH = [12,13,15,12,11,12,11]
>>> parVelocityKMPH = sc.parallelize(airVelocityKMPH,2)
```

The parVelocityKMPH variable is an RDD.

Step 4-5-2. Getting the Number of Data Points

The number of data points gives us an idea of the data size. We apply the count() function to get the number of elements in the RDD:

```
>>>countValue =  parVelocityKMPH.count()
```

Here is the output:

7

The total number of data points is seven.

Step 4-5-3. Summing Air Velocities in a Day

Let's get the summation by using the sum() method:

```
>>>sumValue =  parVelocityKMPH.sum()
```

Here is the output:

86

Step 4-5-4. Finding the Mean Air Velocity

Figure 4-5 shows the mathematical formula for finding a mean, where x1, x2, . . . xn are *n* data points.

$$mean = \frac{\sum_{i=1}^{n} x_i}{n}$$

Figure 4-5. *Calculating the mean*

We calculate the mean by using the mean() function defined on the RDD:

```
>>>meanValue = parVelocityKMPH.mean()
```

Here is the output:

```
12.285714285714286
```

Step 4-5-5. Finding the Variance of Air Data

If we have the data points x1, x2, ... xn, then Figure 4-6 shows the mathematical formula for calculating variance. We are going to calculate the variance of the given air data by using the variance() function defined on the RDD.

```
>>> varianceValue = parVelocityKMPH.variance()
```

$$Variance = \frac{\sum_{i=1}^{n}(x_i - mean)^2}{n}$$

Figure 4-6. *Calculating the variance*

Here is the output:

```
1.63265306122449
```

Step 4-5-6. Calculating Sample Variance

The variance function calculates the population variance. In order to calculate the sample variance, we have to use sampleVariance() defined on the RDD.

For data points x1, x2, ... xn, the sample standard variance is defined in Figure 4-7.

$$sample\,Variance = \frac{\sum_{i=1}^{n}(x_i - mean)^2}{n-1}$$

Figure 4-7. *Calculating the sample variance*

The following line of code calculates the sample standard deviation:

```
>>> sampleVarianceValue = parVelocityKMPH.sampleVariance()
```

Here is the output:

```
1.904761904761905
```

Step 4-5-7. Calculating Standard Deviation

The *standard deviation* is the square root of the variance value. Let's calculate the standard deviation by using the stdev() function:

```
>>> stdevValue = parVelocityKMPH.stdev()
```

Here is the output:

```
1.2777531299998799
```

The standard deviation of the given air velocity data is 1.2777531299998799.

Step 4-5-8. Calculating Sample Standard Deviation

```
>>> sampleStdevValue = parVelocityKMPH.sampleStdev()
```

Here is the output:

```
1.3801311186847085
```

Step 4-5-9. Calculating All Values in One Step

We can calculate all the values of the summary statistics in one go by using the stats() function. The StatCounter object is returned from the stats() function. Let's use the stats() function to calculate the summary statistics of the air velocity data:

```
>>> type(parVelocityKMPH.stats())
```

Here is the output:

```
<class 'pyspark.statcounter.StatCounter'>
```

```
>>> parVelocityKMPH.stats()
```

Here is the output:

```
(count: 7, mean: 12.2857142857, stdev: 1.27775313, max: 15.0, min: 11.0)
```

We can see that the stats() function is an action. It calculates the count, mean, standard deviation, maximum, and minimum of an RDD in one go. It returns the result as a tuple with elements that are key/value pairs. The result of the stats() function can be transformed into a dictionary by using the asDict() function:

```
>>> parVelocityKMPH.stats().asDict()
```

Here is the output:

```
{'count': 7, 'min': 11.0, 'max': 15.0, 'sum': 86.0, 'stdev': 1.3801311186847085,
'variance': 1.904761904761905, 'mean': 12.285714285714286}
```

We also can get individual elements by using different functions defined on StatCounter. Let's start with fetching the mean. The mean value can be found by using the mean() function defined on the StatCounter object:

```
>>> parVelocityKMPH.stats().mean()
```

Here is the output:

```
12.285714285714286
```

Similarly, we can get the number of elements in the RDD, the minimum value, the maximum value, and the standard deviation by using the functions count(), min(), max(), and stdev() functions, respectively. Let's start with the standard deviation:

```
>>> parVelocityKMPH.stats().stdev()
```

Here is the output:

```
1.2777531299998799
```

This command provides the number of elements:

```
>>> parVelocityKMPH.stats().count()
```

Here is the output:

```
7
```

Next, we find the minimum value:

```
>>> parVelocityKMPH.stats().min()
```

Here is the output:

```
11.0
```

Then we find the maximum value:

```
>>> parVelocityKMPH.stats().max()
```

Here is the output:

```
15.0
```

Recipe 4-6. Start PySpark Shell on Standalone Cluster Manager
Problem

You want to start PySpark shell using Standalone Cluster Manager.

Solution

Standalone Cluster Manager provides a master/slave structure. In order to start Standalone Cluster Manager on a single machine, we can use the start-all.sh script. We can find this script inside $SPARK_HOME/sbin. In our case, $SPARK_HOME is /allPySpark/spark. In a cluster environment, we can run a master script to run the master on one machine, and a slave script on a different slave machine to start the slave.

We can start the PySpark shell by using the master URL of the Standalone master. The Standalone master URL looks like spark://<masterHostname>:<masterPort>. It can be easily found in the log file of the master.

How It Works

We have to start the Standalone master and slave processes. There are two ways to do this. In the spark/sbin directory, we will find the start-all.sh script. It will start the master and slave together on the same machine.

Another way to start the processes is as follows: on the master machine, start the master by using the start-master.sh script; and start the slave on a different machine by using start-slave.sh.

Step 4-6-1. Starting Standalone Cluster Manager Using the start-all.sh Script

In this step, we are going to run the start-all.sh script:

```
[pysparkbook@localhost ~]$ /allPySpark/spark/sbin/start-all.sh
```

Here is the output:

```
starting org.apache.spark.deploy.master.Master, logging to /allPySpark/
logSpark//spark-pysparkbook-org.apache.spark.deploy.master.Master-1-
localhost.localdomain.out
localhost: starting org.apache.spark.deploy.worker.Worker, logging to /
allPySpark/logSpark//spark-pysparkbook-org.apache.spark.deploy.worker.
Worker-1-localhost.localdomain.out
```

The logs of starting the Standalone cluster by using the start-all.sh script are written to two files. The file spark-pysparkbook-org.apache.spark.deploy.master.Master-1-localhost.localdomain.out is the log of starting the master process. In order to connect to the master, we need the master URL. We can find this URL in the log file.

Let's open the master log file:

```
[pysparkbook@localhost ~]$ vim /allPySpark/logSpark//spark-pysparkbook-org.
apache.spark.deploy.master.Master-1-localhost.localdomain.out
```

Here is the output:

```
17/03/02 03:33:59 INFO util.Utils: Successfully started service
'sparkMaster' on port 7077.
17/03/02 03:33:59 INFO master.Master: Starting Spark master at spark://
localhost.localdomain:7077
17/03/02 03:33:59 INFO master.Master: Running Spark version 1.6.2
17/03/02 03:34:10 INFO server.Server: jetty-8.y.z-SNAPSHOT
17/03/02 03:34:10 INFO server.AbstractConnector: Started
SelectChannelConnector@0.0.0.0:8080
17/03/02 03:34:10 INFO util.Utils: Successfully started service 'MasterUI'
on port 8080.
17/03/02 03:34:10 INFO ui.MasterWebUI: Started MasterWebUI at
http://10.0.2.15:8080
```

The logs in the master log file indicate that the master has been started on spark://localhost.localdomain:7077 and that the master web UI is at http://10.0.2.15:8080.

Let's open the master web UI, shown in Figure 4-8. In this UI, we can see the URL of the Standalone master.

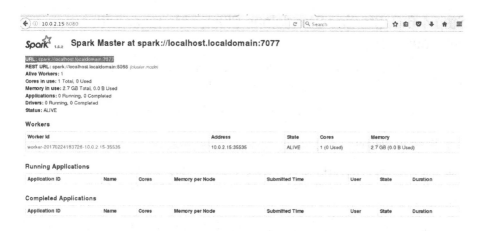

Figure 4-8. *Spark master*

The following command will start the PySpark shell on the Standalone cluster:

```
[pysparkbook@localhost ~]$/allPySpark/spark/bin/pyspark --master spark://
localhost.localdomain:7077
```

After running jps, we can see the master and worker processes:

```
[pysparkbook@localhost ~]$ jps
```

Here is the output:

```
4246 Worker
3977 Master
4285 Jps
```

We can stop Standalone Cluster Manager by using the stop-all.sh script:

```
[pysparkbook@localhost ~]$ /allPySpark/spark/sbin/stop-all.sh
```

Here is the output:

```
localhost: stopping org.apache.spark.deploy.worker.Worker
stopping org.apache.spark.deploy.master.Master
```

We can see that the stop-all.sh script first stops the worker (slave) and then stops the master.

Step 4-6-2. Starting Standalone Cluster Manager with an Individual Script

The PySpark framework provides an individual script for starting the master and workers on different machines. Let's first start the master by using the start-master.sh script:

```
[pysparkbook@localhost ~]$ /allPySpark/spark/sbin/start-master.sh
```

Here is the output:

```
starting org.apache.spark.deploy.master.Master, logging to /allPySpark/
logSpark//spark-pysparkbook-org.apache.spark.deploy.master.Master-1-
localhost.localdomain.out
```

Again, as done previously, we have to get the master URL from the log file. Using that master URL, we can start slaves on different machines by using the start-slave.sh script:

```
[pysparkbook@localhost ~]$ /allPySpark/spark/sbin/start-slave.sh spark://
localhost.localdomain:7077
```

Here is the output:

```
starting org.apache.spark.deploy.worker.Worker, logging to /allPySpark/
logSpark//spark-pysparkbook-org.apache.spark.deploy.worker.Worker-1-
localhost.localdomain.out
```

Using the jps command, we can see the worker and master processes running on our machine:

```
[pysparkbook@localhost ~]$ jps
```

Here is the output:

```
4246 Worker
3977 Master
4285 Jps
```

You might remember that when we use stop-all.sh, the script first stops the worker process and then the master process. We have to follow this order. We stop the worker process first, followed by the master process.

To stop the worker process, use the stop-slave.sh script:

```
[pysparkbook@localhost ~]$ /allPySpark/spark/sbin/stop-slave.sh
```

Here is the output:

```
stopping org.apache.spark.deploy.worker.Worker
```

Running the jps command will show only the master process now:

```
[pysparkbook@localhost ~]$ jps
```

Here is the output:

```
4341 Jps
3977 Master
```

Similarly, we can stop the master process by using the stop-master.sh script:

```
[pysparkbook@localhost ~]$ /allPySpark/spark/sbin/stop-master.sh
```

Here is the output:

```
stopping org.apache.spark.deploy.master.Master
```

Recipe 4-7. Start PySpark Shell on Mesos
Problem

You want to start PySpark shell on a Mesos cluster manager.

Solution

Mesos is another cluster manager. Mesos also follows a master/slave architecture, similar to Standalone Cluster Manager. In order to start Mesos, we have to first start the Mesos master and then the Mesos slave. Then the PySpark shell can be started by using the master URL on Mesos.

How It Works

We installed Mesos in Chapter 3. Now we have to start the master and slave processes one by one. After starting the master and slaves, we have to start the PySpark shell. The following command starts the Mesos master process:

```
[pysparkbook@localhost ~]$ mesos-master –work_dir=/allPySpark/mesos/workdir
```

Here is the output:

```
I0224 09:50:57.575908  9839 main.cpp:263] Build: 2016-12-29 00:42:08 by
pysparkbook

I0224 09:50:57.576501  9839 main.cpp:264] Version: 1.1.0
I0224 09:50:57.582787  9839 main.cpp:370] Using 'HierarchicalDRF' allocator
```

The slave process will start subsequent to the start of the master process:

```
[root@localhost binaries]#mesos-slave --master=127.0.0.1:5050 --work_dir=/
allPySpark/mesos/workdir1 --systemd_runtime_directory=/allPySpark/mesos/systemd
```

Here is the output:

```
I0224 18:22:25.002970  3797 gc.cpp:55] Scheduling '/allPySpark/mesos/
workdirSlave/slaves/dd2c5f22-57f9-416e-a71a-0cc83de8558d-S1/frameworks/
dd2c5f22-57f9-416e-a71a-0cc83de8558d-0000/executors/1/runs/a7c3d613-9696-
4d42-afe2-a27c5c825e72' for gc 6.99998839417778days in the future
I0224 18:22:25.003083  3797 gc.cpp:55] Scheduling
I0224 18:22:25.003123  3797 gc.cpp:55] Scheduling '/allPySpark/mesos/
workdirSlave/slaves/dd2c5f22-57f9-416e-a71a-0cc83de8558d-S1/frameworks/
dd2c5f22-57f9-416e-a71a-0cc83de8558d-0000' for gc 6.99998839205926days in
the future
```

We have our master and slave processes started. Now we can start the PySpark shell by using the following command. We provide the master URL and one parameter, spark.executor.uri. The spark.executor.uri parameter tells Mesos the location to get the PySpark assembly.

```
[pysparkbook@localhost binaries]$ /allPySpark/spark/bin/pyspark --master
mesos://127.0.0.1:5050 --conf spark.executor.uri=/home/pysparkbook/binaries/
spark-2.0.0-bin-hadoop2.6.tgz
```

We can run the jps command to see the Java process after running PySpark on Mesos:

```
[pysparkbook@localhost binaries]$jps
```

Here is the output:

```
4174 SparkSubmit
4287 CoarseGrainedExecutorBackend
```

■ ■ ■

CHAPTER 5

The Power of Pairs: Paired RDDs

Key/value pairs are good for solving many problems efficiently in a parallel fashion. Apache Mahout, a machine-learning library that was initially developed on top of Apache Hadoop, implements many machine-learning algorithms in the areas of classification, clustering, and collaborative filtering by using the MapReduce key/value-pair architecture. In this chapter, you'll work through recipes that develop skills for solving interesting big data problems from many disciplines.

This chapter covers the following recipes:

> Recipe 5-1. Create a paired RDD
>
> Recipe 5-2. Perform data aggregation
>
> Recipe 5-3. Join data
>
> Recipe 5-4. Calculate page rank

Recipe 5-1. Create a Paired RDD
Problem

You want to create a paired RDD.

Solution

You have an RDD, RDD1. The elements of RDD1 are b, d, m, t, e, and u. You want to create a paired RDD, in which the keys are elements of a single RDD, and the value of a key is 0 if the element is a consonant, or 1 if the element is a vowel. Figure 5-1 clearly depicts the requirements.

© Raju Kumar Mishra 2018
R. K. Mishra, *PySpark Recipes*, https://doi.org/10.1007/978-1-4842-3141-8_5

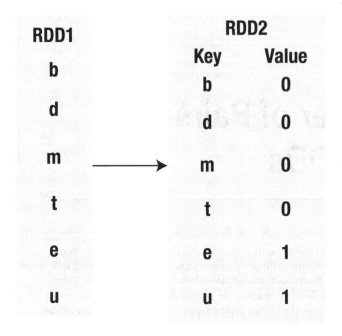

Figure 5-1. *Creating a paired RDD*

A paired RDD can be created in many ways. One way is to read data directly from files. We'll explore this method in an upcoming chapter. Another way to create a paired RDD is by using the map() method, which you'll learn about in this recipe.

How It Works

In this section, you'll follow several steps to reach the solution.

Step 5-1-1. Creating an RDD with Single Elements

Let's start by creating an RDD out of our given data:

```
>>> pythonList = ['b' , 'd', 'm', 't', 'e', 'u']
>>> RDD1 = sc.parallelize(pythonList,2)
>>>RDD1.collect()
```

Here is the output:

```
['b', 'd', 'm', 't', 'e', 'u']
```

We have created an RDD named RDD1. The elements of RDD1 are b, d, m, t, e, and u. This is an RDD of letters. It can be observed that the elements b, d, m, and t are consonants. The other elements of RDD1, e and u, are vowels.

Step 5-1-2. Writing a Python Method to Check for Consonants

We are going to define a Python function named vowelCheckFunction(). This function will take a letter as input and return 1 if the input is a consonant, or 0 if it is not. Let's implement the function:

```
>>> def vowelCheckFunction( data) :
...     if data in ['a','e','i','o','u']:
...         return 1
...     else :
...         return 0
```

It can be observed that our Python function vowelCheckFunction will take one input data. In this case, we are going to send a letter. Inside the function, we check whether our data is a vowel. If our if block logical expression results in True, our function will return 1; otherwise, it will return 0.

Without testing vowelCheckFunction, we shouldn't trust it. So let's start with a vowel as input:

```
>>> vowelCheckFunction('a')
```

Here is the output:

```
1
```

We get 1 as the output for our vowel input. Our Python function vowelCheckFunction meets our expectation for vowels. That's nice, but let's test for consonants too. This time we are going to send b, a consonant, as the input:

```
>>> vowelCheckFunction('b')
```

Here is the output:

```
0
```

We have tested our function. It is working for both consonants and vowels with the anticipated output. We can bank on our developed function.

Step 5-1-3. Creating a Paired RDD

We can create our required RDD by using the map() function. We have to create a paired RDD: the keys will be the elements of RDD1, and the value will be 0 for keys that are consonants, or 1 for keys that are vowels:

```
>>> RDD2 = RDD1.map( lambda data : (data, vowelCheckFunction(data)))

>>>RDD2.collect()
```

Here is the output:

```
[('b', 0),
 ('d', 0),
 ('m', 0),
 ('t', 0),
 ('e', 1),
 ('u', 1)]
```

Step 5-1-4. Fetching Keys from a Paired RDD

The keys() function can be used to fetch all the keys:

```
>>> RDD2Keys = RDD2.keys()
The following code line gets all the keys:
>>> RDD2Keys.collect()
```

Here is the output:

```
['b', 'd', 'm', 't', 'e', 'u']
```

We can see that the keys() function performs a transformation. Therefore, keys() returns an RDD that requires the collect() function to get the data to the driver.

Step 5-1-5. Fetching Values from a Paired RDD

Similar to the keys() function, the values() function will fetch all the values from a paired RDD. It also performs a transformation:

```
>>> RDD2Values = RDD2.values()
>>> RDD2Values.collect()
```

Here is the output:

```
[0, 0, 0, 0, 1, 1]
```

Recipe 5-2. Aggregate Data

Problem

You want to aggregate data.

Solution

You want to perform data aggregation on data from a lightbulb manufacturer, as shown in Table 5-1.

Table 5-1. *Filament Data*

Filament Type	Bulb Power	Life in Hours
filamentA	100W	605
filamentB	100W	683
filamentB	100W	691
filamentB	200W	561
filamentA	200W	530
filamentA	100W	619
filamentB	100W	686
filamentB	200W	600
filamentB	100W	696
filamentA	200W	579
filamentA	200W	520
filamentA	100W	622
filamentA	100W	668
filamentB	200W	569
filamentB	200W	555
filamentA	200W	541

Company YP manufactures two types of filaments: filamentA and filamentB. 100W and 200W electric bulbs can be manufactured from both filaments. Table 5-1 indicates the expected life of each bulb.

You want to calculate the following:

- Mean life in hours for bulbs of each filament type

- Mean life in hours for bulbs of each power level

- Mean life in hours based on both filament type and power

We generally encounter aggregation of data in data-science problems. To get an aggregation of data, we can use many PySpark functions.

In this recipe, we'll use the reduceByKey() function to calculate the mean by using keys. Calculating the mean of complex keys requires creating those complex keys. Complex keys can be created by using the map() function.

How It Works

Let's start with the creation of the RDD.

Step 5-2-1. Creating an RDD with Single Elements

```
>>> filDataSingle = [['filamentA','100W',605],
...                  ['filamentB','100W',683],
...                  ['filamentB','100W',691],
...                  ['filamentB','200W',561],
...                  ['filamentA','200W',530],
...                  ['filamentA','100W',619],
...                  ['filamentB','100W',686],
...                  ['filamentB','200W',600],
...                  ['filamentB','100W',696],
...                  ['filamentA','200W',579],
...                  ['filamentA','200W',520],
...                  ['filamentA','100W',622],
...                  ['filamentA','100W',668],
...                  ['filamentB','200W',569],
...                  ['filamentB','200W',555],
...                  ['filamentA','200W',541]]

>>> filDataSingleRDD = sc.parallelize(filDataSingle,2)

>>> filDataSingleRDD.take(3)
```

Here is the output:

```
[['filamentA', '100W', 605],
 ['filamentB', '100W', 683],
 ['filamentB', '100W', 691]]
```

Here we are first creating a nested Python list of filament data named filDataSingle. Then we create the RDD filDataSingleRDD. We divide our data into two parts. The output of the take() function on filDataSingleRDD clearly shows that the elements of the RDD are a list.

Step 5-2-2. Creating a Paired RDD

First we have to calculate the mean lifetime of bulbs, based on their filament type. Better that we are creating a paired RDD with keys for the filament type and values for the life in hours. So let's create our required paired RDD and then investigate it:

```
>>> filDataPairedRDD1 = filDataSingleRDD.map(lambda x : (x[0], x[2]))
>>> filDataPairedRDD1.take(4)
```

Here is the output:

```
[('filamentA', 605),
 ('filamentB', 683),
 ('filamentB', 691),
 ('filamentB', 561)]
```

We have created a paired RDD, filDataPairedRDD1, by using the map() function defined on the RDD. The paired RDD filDataPairedRDD1 has the filament type as the key, and the life in hours as the value.

Step 5-2-3. Finding the Mean Lifetime Based on Filament Type

Now we have our required paired RDD. But is this all we need? No. To calculate the mean, we need a sum and a count. We have to add an extra 1 in our paired RDD so that we can get a sum and a count. So let's add an extra 1 now to each RDD element:

```
>>> filDataPairedRDD11 = filDataPairedRDD1.map(lambda x : (x[0], [x[1], 1]))
>>> filDataPairedRDD11.take(4)
```

Here is the output:

```
[('filamentA', [605, 1]),
 ('filamentB', [683, 1]),
 ('filamentB', [691, 1]),
 ('filamentB', [561, 1])]
```

filDataPairedRDD11 is a paired RDD. The values of filDataPairedRDD11 are presented as a list; the first element is the lifetime of the bulb (in hours), and the second element is just 1.

Now we have to calculate the sum of the values of the lifetimes for each filament type as well as the count value, so that we can calculate the mean. Many PySpark functions could be used to do this job, but here we are going to use the reduceByKey() function for paired RDDs.

The reduceByKey() function applies aggregation operators key wise. It takes an aggregation function as input and applies that function on the values of each RDD key.

Let's calculate the sum of the total life hours of bulbs based on the filament type, and the count of elements for each filament type:

```
>>> filDataSumandCount = filDataPairedRDD11.reduceByKey(lambda l1,l2 :
[l1[0] + l2[0] ,l1[1]+l2[1]])

>>> filDataSumandCount.collect()
```

Here is the output:

```
[('filamentA', [4684, 8]),
 ('filamentB', [5041, 8])]
```

We are applying the reduceBykey() function to our paired RDD, filDataPairedRDD11. To understand the workings of reduceByKey(), let's see how our paired RDD has been distributed. Let's start with the number of elements in the RDD: the five elements of filDataPairedRDD11:

```
>>> filDataPairedRDD11.count()
```

Here is the output:

```
16
```

Our paired RDD has 16 elements. Now let's see how many partitions our data has:

```
>>> filDataPairedRDD11.getNumPartitions()
```

Here is the output:

```
2
```

It is clear that the data has been distributed in two parts. PySpark will try to distribute data evenly to executors, so it will distribute eight data points to each executor. Now let's take five data points out of our RDD and see what that data looks like:

```
>>> filDataPairedRDD11.take(5)
```

Here is the output:

```
[('filamentA', [605, 1]),
 ('filamentB', [683, 1]),
 ('filamentB', [691, 1]),
 ('filamentB', [561, 1]),
 ('filamentA', [530, 1])]
```

We can see that first and fifth element of our filDataPairedRDD11 RDD has the key filamentA. Therefore, the first l1 and l2 of our reduceByKey() function for filament type filamentA will be [605, 1] and [530, 1], respectively.

```
>>> filDataSumandCount.collect()
```

Here is the output:

```
[('filamentA', [4684, 8]),
 ('filamentB', [5041, 8])]
```

Finally, we have the summation of the life hours of bulbs and the count, based on filament type. The next step is to divide the sum by the count to get the mean value. Let's do that:

```
>>> filDataMeanandCount = filDataSumandCount.map( lambda l :
[l[0],float(l[1][0])/l[1][1],l[1][1]])
```

```
>>> filDataMeanandCount.collect()
```

Here is the output:

```
[['filamentA', 585.5, 8],
 ['filamentB', 630.125, 8]]
```

Finally, we have our required mean, based on filament type. The mean lifetime of filamentA is 585.5 hours, and the mean lifetime of filamentB is 630.125 hours. We can infer that filamentB has a longer life than filamentA.

Step 5-2-4. Finding the Mean Lifetime Based on Bulb Power

First, we will start with creating our paired RDD. The key will be the bulb power, and the value will be the life in hours:

```
>>> filDataPairedRDD2 = filDataSingleRDD.map(lambda x : (x[1], x[2]))
```

```
>>> filDataPairedRDD2.take(4)
```

Here is the output:

```
[('100W', 605),
 ('100W', 683),
 ('100W', 691),
 ('200W', 561)]
```

We have created a paired RDD, filDataPairedRDD2, and each element is a pair: of bulb power and the corresponding life in hours.

```
>>> fillDataPairedRDD22 = filDataPairedRDD2.map( lambda x : (x[0],[x[1],1]))

>>> fillDataPairedRDD22.take(4)
```

Here is the output:

```
[('100W', [605, 1]),
 ('100W', [683, 1]),
 ('100W', [691, 1]),
 ('200W', [561, 1])]
```

Now we have included 1 in the value part of the RDD. Therefore, each value is a list that consists of the life in hours and a 1.

```
>>> powerSumandCount = fillDataPairedRDD22.reduceByKey(lambda l1,l2 :
[l1[0]+l2[0], l1[1]+l2[1]])

>>> powerSumandCount.collect()
```

Here is the output:

```
[('100W', [5270, 8]),
 ('200W', [4455, 8])]
```

We have calculated the sum of the total life hours and the count, with bulb power as the key.

```
>>> meanandCountPowerWise =powerSumandCount.map(lambda val :
[val[0],[float(val[1][0])/val[1][1],val[1][1]]])

>>> meanandCountPowerWise.collect()
```

Here is the output:

```
[['100W', [658.75, 8]], ['200W', [556.875, 8]]]
```

In this last step, we have computed the mean and the count. From the result, we can infer that the mean life of 100W bulbs is longer than that of 200W bulbs.

Step 5-2-5. Finding the Mean Lifetime Based on Filament Type and Power

To solve this part of the exercise, we need a paired RDD with keys that are complex. You might be wondering what a complex key is. *Complex keys* have more than one type. In our case, our complex key will have both the filament type and bulb power type. Let's start creating our paired RDD with a complex key type:

```
>>> filDataSingleRDD.take(4)
```

Here is the output:

```
[['filamentA', '100W', 605],
 ['filamentB', '100W', 683],
 ['filamentB', '100W', 691],
 ['filamentB', '200W', 561]]
```

```
>>> filDataComplexKeyData = filDataSingleRDD.map( lambda val : [(val[0],
val[1]),val[2]])
```

```
>>> filDataComplexKeyData.take(4)
```

Here is the output:

```
[[('filamentA', '100W'), 605],
 [('filamentB', '100W'), 683],
 [('filamentB', '100W'), 691],
 [('filamentB', '200W'), 561]]
```

We have created a paired RDD named filDataComplexKeyData. It can be easily observed that it has complex keys. The keys are a combination of filament type and bulb power. The rest of the exercise will move as in the previous step. In the following code, we are going to include an extra 1 in the values:

```
>>> filDataComplexKeyData1 = filDataComplexKeyData.map(lambda val : [val[0]
,[val[1],1]])
```

```
>>> filDataComplexKeyData1.take(4)
```

Here is the output:

```
[[('filamentA', '100W'), [605, 1]],
 [('filamentB', '100W'), [683, 1]],
 [('filamentB', '100W'), [691, 1]],
 [('filamentB', '200W'), [561, 1]]]
```

125

Our required paired RDD, filDataComplexKeyData1, has been created. Now we can apply the reduceByKey() function to get the sum and count, based on the complex keys:

```
>>> filDataComplexKeySumCount = filDataComplexKeyData1.reduceByKey(lambda
l1,l2 : [l1[0]+l2[0], l1[1]+l2[1]])

>>> filDataComplexKeySumCount.collect()
```

Here is the output:

```
[(('filamentA', '100W'), [2514, 4]),
 (('filamentB', '200W'), [2285, 4]),
 (('filamentB', '100W'), [2756, 4]),
 (('filamentA', '200W'), [2170, 4])]
```

After getting the sum, the mean can be calculated as follows:

```
>>> filDataComplexKeyMeanCount = filDataComplexKeySumCount.map(lambda val :
[val[0],[float(val[1][0])/val[1][1],val[1][1]]])

>>> filDataComplexKeyMeanCount.collect()
```

```
[[('filamentA', '100W'), [628.5, 4]],
 [('filamentB', '200W'), [571.25, 4]],
 [('filamentB', '100W'), [689.0, 4]],
 [('filamentA', '200W'), [542.5, 4]]]
```

Finally, we have calculated the mean and the count value.

■ **Note**   The following is a good tutorial about working with

reduceByKey(): http://stackoverflow.com/questions/30145329/reducebykey-how-does-it-work-internally.

Recipe 5-3. Join Data
Problem

You want to join data.

Solution

We have been given two tables: a Students table (Table 5-2) and a Subjects table (Table 5-3).

Table 5-2. *Students*

Student ID	Name	Gender
si1	Robin	M
si2	Maria	F
si3	Julie	F
si4	Bob	M
si6	William	M

Table 5-3. *Subjects*

Student ID	Subjects
si1	Python
si3	Java
si1	Java
si2	Python
si3	Ruby
si4	C++
si5	C
si4	Python
si2	Java

You want to perform the following on the Students and Subjects tables:

- Inner join

- Left outer join

- Right outer join

- Full outer join

Joining data tables is an integral part of data preprocessing. We are going to perform four types of data joins in this recipe.

An *inner join* returns all the keys that are common to both tables. It discards the key elements that are not common to both tables. In PySpark, an inner join is done by using the join() method defined on the RDD.

A *left outer join* includes all keys in the left table and excludes uncommon keys from the right table. A left outer join can be performed by using the leftOuterJoin() function defined on the RDD in PySpark.

Another important type of join is a *right outer join*. In a right outer join, every key of the second table is included, but from the first table, only those keys that are common to both tables are included. We can do a right outer join by using the rightOuterJoin() function in PySpark.

If you want to include all keys from both tables, go for a *full outer join*. It can be performed by using fullOuterJoin().

How It Works

We'll follow the steps in this section to work with joins.

Step 5-3-1. Creating Nested Lists

Let's start creating a nested list of our data from the Students table:

```
>>> studentData = [['si1','Robin','M'],
...                ['si2','Maria','F'],
...                ['si3','Julie','F'],
...                ['si4','Bob',  'M'],
...                ['si6','William','M']]
```

After creating our Students table data, the next step is to create a nested list of Subjects table data:

```
>>> subjectsData = [['si1','Python'],
...                 ['si3','Java'],
...                 ['si1','Java'],
...                 ['si2','Python'],
...                 ['si3','Ruby'],
...                 ['si4','C++'],
...                 ['si5','C'],
...                 ['si4','Python'],
...                 ['si2','Java']]
```

We have created nested lists from the Students table and Subjects table.

Step 5-3-2. Creating a Paired RDD of Students and Subjects

Before creating a paired RDD, we first have to create a single RDD. Let's create studentRDD:

```
>>> studentRDD = sc.parallelize(studentData, 2)
```

```
>>> studentRDD.take(4)
```

```
[['si1', 'Robin', 'M'],
 ['si2', 'Maria', 'F'],
 ['si3', 'Julie', 'F'],
 ['si4', 'Bob', 'M']]
```

We can see that, every element of the studentRDD RDD is a list, and each list has three elements. Now we have to transform it into a paired RDD:

```
>>> studentPairedRDD = studentRDD.map(lambda val : (val[0],[val[1],val[2]]))

>>> studentPairedRDD.take(4)
```

Here is the output:

```
[('si1', ['Robin', 'M']),
 ('si2', ['Maria', 'F']),
 ('si3', ['Julie', 'F']),
 ('si4', ['Bob', 'M'])]
```

The paired RDD, studentPairedRDD, has the student ID as the key. Now we have to create a paired RDD of the subject data:

```
>>> subjectsPairedRDD = sc.parallelize(subjectsData, 2)

>>> subjectsPairedRDD.take(4)
```

Here is the output:

```
[['si1', 'Python'],
 ['si3', 'Java'],
 ['si1', 'Java'],
 ['si2', 'Python']]
```

We do not need to do anything extra to create a paired RDD of subject data.

Step 5-3-3. Performing an Inner Join

As we know, an inner join in PySpark is done by using the join() function. We have to apply this function on the paired RDD studentPairedRDD, and provide subjectsPairedRDD as an argument to the join() function:

```
>>> studenSubjectsInnerJoin = studentPairedRDD.join(subjectsPairedRDD)

>>> studenSubjectsInnerJoin.collect()
```

Here is the output:

```
[('si3', (['Julie', 'F'], 'Java')),
 ('si3', (['Julie', 'F'], 'Ruby')),
 ('si2', (['Maria', 'F'], 'Python')),
 ('si2', (['Maria', 'F'], 'Java')),
 ('si1', (['Robin', 'M'], 'Python')),
 ('si1', (['Robin', 'M'], 'Java')),
 ('si4', (['Bob', 'M'], 'C++')),
 ('si4', (['Bob', 'M'], 'Python'))]
```

Analyzing the output of this inner join reveals that the key part contains only keys that are common to the Students and Subjects tables; these appear in the joined table. The keys that are not common to both tables are not the part of joined table.

Step 5-3-4. Performing a Left Outer Join

A left outer join can be performed by using the leftOuterJoin() function:

```
>>> studentSubjectsleftOuterJoin = studentPairedRDD.leftOuterJoin(subjects
PairedRDD)

>>> studentSubjectsleftOuterJoin.collect()
```

Here is the output:

```
[('si3', (['Julie', 'F'], 'Java')),
 ('si3', (['Julie', 'F'], 'Ruby')),
 ('si2', (['Maria', 'F'], 'Python')),
 ('si2', (['Maria', 'F'], 'Java')),
 ('si6', (['William', 'M'], None)),
 ('si1', (['Robin', 'M'], 'Python')),
 ('si1', (['Robin', 'M'], 'Java')),
 ('si4', (['Bob', 'M'], 'C++')),
 ('si4', (['Bob', 'M'], 'Python'))]
```

Student ID si6 is in the Students table but not in the Subjects table. Hence, the left outer join includes si6 in the joined table. Because si6 doesn't have its counterpart in the Subjects table, it has None in place of the subject.

Step 5-3-5. Performing a Right Outer Join

A right outer join on the Students and Subjects tables can be performed by using the rightOuterJoin() function:

```
>>> studentSubjectsrightOuterJoin = studentPairedRDD.rightOuterJoin(subjects
PairedRDD)

>>> studentSubjectsrightOuterJoin.collect()
```

Here is the output:

```
[('si3', (['Julie', 'F'], 'Java')),
 ('si3', (['Julie', 'F'], 'Ruby')),
 ('si2', (['Maria', 'F'], 'Python')),
 ('si2', (['Maria', 'F'], 'Java')),
 ('si1', (['Robin', 'M'], 'Python')),
 ('si1', (['Robin', 'M'], 'Java')),
 ('si5', (None, 'C')),
 ('si4', (['Bob', 'M'], 'C++')),
 ('si4', (['Bob', 'M'], 'Python'))]
```

Student ID si5 is in only the Subjects table; it is not part of the Students table. Therefore, it appears in the joined table.

Step 5-3-6. Performing a Full Outer Join

Now let's perform a full outer join. In a full outer join, keys from both tables will be included:

```
>>> studentSubjectsfullOuterJoin = studentPairedRDD.fullOuterJoin(subjects
PairedRDD)

>>> studentSubjectsfullOuterJoin.collect()
```

Here is the output:

```
[('si3', (['Julie', 'F'], 'Java')),
 ('si3', (['Julie', 'F'], 'Ruby')),
 ('si2', (['Maria', 'F'], 'Python')),
 ('si2', (['Maria', 'F'], 'Java')),
 ('si6', (['William', 'M'], None)),
 ('si1', (['Robin', 'M'], 'Python')),
 ('si1', (['Robin', 'M'], 'Java')),
 ('si5', (None, 'C')),
 ('si4', (['Bob', 'M'], 'C++')),
 ('si4', (['Bob', 'M'], 'Python'))]
```

In the joined table, keys from both tables have been included. Student ID si6 is part of only the Students data, and it appears in the joined table. Similarly, student ID si5 is part of only the Subjects table, but it appears in our joined table.

Recipe 5-4. Calculate Page Rank
Problem

You want to calculate the page rank of the web-page system illustrated in Figure 5-2.

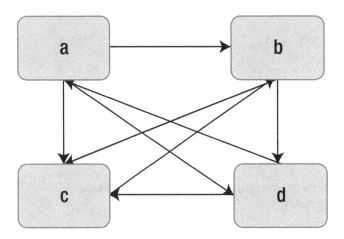

Figure 5-2. *A web-page system*

We have four web pages (a, b, c, and d) in our system. Web page a has outbound links to pages b, c, and d. Similarly, page b has an outbound link to pages d and c. Web page c has an outbound link to page b, and page d has an outbound link to pages a and c.

Solution

The *page-rank algorithm* was developed by Sergey Brin and Larry Page. In the algorithm name, *page* stands for *Larry Page*. The developers of the page-rank algorithm later founded Google. It is an iterative algorithm.

The page rank of a particular web page indicates its relative importance within a group of web pages. The higher the page rank, the higher up it will appear in a search result list.

The importance of a page is defined by the importance of all the web pages that provide an outbound link to the web page in consideration. For example, say that web page X has very high relative importance. Web page X is outbounding to web page Y; hence, web page Y will also have high importance.

Let's summarize the page-rank algorithm:

1. Initialize each page with a page rank of 1 or some arbitrary number.

2. For i in someSequence, do the following:

   a. Calculate the contribution of each inbound page.

   b. Calculate the sum of the contributions.

   c. Determine the updated page rank as follows:

   ```
   updated page rank = 1 - s + s × summation of contributions
   ```

Let me show you an example. Say PR(a), PR(b), PR(c), and PR(d) are the page ranks of pages a, b, c, and d, respectively. Page d has two inbound links: the first from page a, and the second from page b.

Now we have to know how the contribution to page rank by a web page is calculated. This contribution to page rank is given by the following formula:

$$\text{Contribution to a page} = \frac{\text{page rank of contributing page}}{\text{total number of outbounds page from the contributing page}}$$

In our example, web page a has three outbound links: the first is to page b, the second is to page c, and the last is to page d. So the contribution to page rank of page d by page a is PR(a) / 3. Now we have to calculate the contribution of page b to page d. Page b has two outbound links: the first to page c, and the second to page d. Hence, the contribution by page b is PR(b) / 2.

So the page rank of page d will be updated as follows, where s is known as the *damping factor*:

$$PR(d) = 1 - s + s \times (PR(a)/3 + PR(b)/2)$$

■ **Note** For more information on page rank, visit Wikipedia at https://en.wikipedia.org/wiki/PageRank.

How It Works

Follow the steps in this section to work with page rank.

Step 5-4-1. Creating Nested Lists of Web Pages with Outbound Links and Initializing Rank

```
>>> pageLinks =  [['a' ,['b','c','d']],
                   ['c', ['b']],

...                          ['b', ['d','c']],
...                          ['d', ['a','c']]]
```

We have created a nested list of web pages and their outbound links. Now we have to initialize the page ranks of all the pages. We are initializing them by 1:

```
>>> pageRanks =  [['a',1],
...               ['c',1],
...               ['b',1],
...               ['d',1]]
```

After creating ranks for the nested list, we have to define the number of iterations for running the page rank.

Step 5-4-2. Writing a Function to Calculate Contributions

We are going to write a function that will take two arguments. The first argument of our function is a list of web page URIs, which provide the outbound links to other web pages. The second argument is the rank of the web page accessed through the outbound links that are the first argument. The function will return the contribution to all the web pages in the first argument:

```
>>> def rankContribution(uris, rank):
...     numberOfUris = len(uris)
...     rankContribution = float(rank) / numberOfUris
...     newrank =[]
...     for uri in uris:
...        newrank.append((uri, rankContribution))
...     return newrank
```

This code is very explicable. Our function, rankContribution, will return the contribution to the page rank for the list of URIs (first variable). The function will first calculate the number of elements in our list URIs; then it will calculate the rank contributions to the given URIs. And finally, for each URI, the contributed rank will be returned.

Step 5-4-3. Creating Paired RDDs

Let's first create our paired RDDs of link data:

```
>>> pageLinksRDD   = sc.parallelize(pageLinks, 2)
>>> pageLinksRDD.collect()
```

Here is the output:

```
[['a', ['b', 'c', 'd']],
 ['c', ['b']],
 ['b', ['d', 'c']],
 ['d', ['a', 'c']]]
```

And then we'll create the paired RDD of our rank data:

```
>>> pageRanksRDD   = sc.parallelize(pageRanks, 2)
>>> pageRanksRDD.collect()
```

Here is the output:

```
[['a', 1],
 ['c', 1],
 ['b', 1],
 ['d', 1]]
```

Step 5-4-4. Creating a Loop for Updating Page Rank

Now it is time to write our final loop to update the page rank of every page:

```
>>>numIter = 20
>>>s = 0.85
```

We have defined the number of iterations and the damping factor, s.

```
>>> for i in range(numIter):
...      linksRank = pageLinksRDD.join(pageRanksRDD)
...      contributedRDD = linksRank.flatMap(lambda x : rankContribution(x[1]
[0],x[1][1]))
...      sumRanks = contributedRDD.reduceByKey(lambda v1,v2 : v1+v2)
...      pageRanksRDD = sumRanks.map(lambda x : (x[0],(1-s)+s*x[1]))
...
```

Let's investigate our for loop. First we join pageLinksRDD to pageRanksRDD via an inner join. Then the second line of the for block calculates the contribution to the page rank by using the rankContribution() function we defined previously.

In the next line, we aggregate all the contributions we have. In the last line, we update the rank of each web page by using the map() function. Now it is time to enjoy the result:

```
>>> pageRanksRDD.collect()
```

Here is the output:

```
[('b', 1.3572437951279825),
 ('c', 1.2463781024360086),
 ('d', 0.8746512999550939),
 ('a', 0.521726802480915)]
```

Finally, we have the estimated page rank for every page.

CHAPTER 6

■ ■ ■

I/O in PySpark

File input/output (I/O) operations are an integral part of many software activities and for data

A data scientist deals with many types of files, including text files, comma-separated values (CSV) files, JavaScript Object Notation (JSON) files, and many more. The Hadoop Distributed File System (HDFS) is a very good distributed file system.

This chapter covers the following recipes:

Recipe 6-1. Read a simple text file

Recipe 6-2. Write an RDD to a simple text file

Recipe 6-3. Read a directory

Recipe 6-4. Read data from HDFS

Recipe 6-5. Save an RDD to HDFS

Recipe 6-6. Read data from a sequential file

Recipe 6-7. Write data into a sequential file

Recipe 6-8. Read a CSV file

Recipe 6-9. Write an RDD to a CSV file

Recipe 6-10. Read a JSON file

Recipe 6-11. Write an RDD to a JSON file

Recipe 6-12. Read table data from HBase by using PySpark

Recipe 6-1. Read a Simple Text File
Problem

You want to read a simple text file.

Solution

You have a simple text file named shakespearePlays.txt. The file content is as follows:

> Love's Labour's Lost
>
> A Midsummer Night's Dream
>
> Much Ado About Nothing
>
> As You Like It

The shakespearePlays.txt file has four lines. You want to read this file by using PySpark. After reading the file, you want to calculate the following:

- Total number of lines in the file

- Total number of characters in the file

To read a simple file, you can use two functions: textFile() and wholeTextFiles(). These two functions are defined on our SparkContext object.

The textFile() method reads a text file and results in an RDD of lines. The textFile() method is a transformation, so textFile() does not read the data until the first action is called. Because the file is not available at the time textFile() is run, it will not throw an error. It will throw an error when the first action is called. Why is this? Like other transformations, the textFile() function will be called when the first action is called.

Another method, wholeTextFiles(), works in similar way as textFile() except it reads the file as a key/value pair. The file name is read as the key, and the file data is read as the value associated with that key.

How It Works

Let's see how these built-in methods work.

Step 6-1-1. Reading a Text File by Using the textFile() Function

The textFile() function takes three inputs. The first input to textFile() is the path of the file that has to be read. The second argument is minPartitions, which defines the minimum number of data partitions in the RDD. The third argument is use_unicode. If use_unicode is False, the file is read as a string. Here is the textFile() function:

```
>>> playData = sc.textFile('/home/pysparkbook/pysparkBookData/
shakespearePlays.txt',2)
```

Here is the output:

```
17/03/18 07:10:17 INFO MemoryStore: Block broadcast_8 stored as values in
memory (estimated size 61.8 KB, free 208.0 KB)
17/03/18 07:10:17 INFO MemoryStore: Block broadcast_8_piece0 stored as bytes
in memory (estimated size 19.3 KB, free 227.3 KB)
```

In the preceding code, we have provided the file path and set minPartitions to 2.

```
>>> playDataList = playData.collect()
>>> type(playDataList)
```

Here is the output:

```
<type 'list'>
```

We can see that the textFile() function has read the file and created the RDD. The elements of the RDD are lines:

```
>>> playDataList[0:4]
```

Here is the output:

```
[u"Love's Labour's Lost",
 u"A Midsummer Night's Dream",
 u'Much Ado About Nothing',
 u'As You Like It']
```

Step 6-1-2. Reading a Text File by Using wholeTextFiles()

The wholeTextFiles() function also takes three inputs. The first input to wholeTextFiles() is the path of the file that has to be read. The second argument is minPartitions, which defines the minimum number of data partitions in the RDD. The third argument is use_unicode. If use_unicode is False, the file is read as a string. Let's read the same text file, this time by using the wholeTextFiles() function:

```
>>> playData = sc.wholeTextFiles('/home/pysparkbook/pysparkBookData/
shakespearePlays.txt',2)
```

Here is the output:

```
17/03/18 07:19:06 INFO MemoryStore: Block broadcast_11 stored as values in
memory (estimated size 209.0 KB, free 446.0 KB)
17/03/18 07:19:06 INFO MemoryStore: Block broadcast_11_piece0 stored as
bytes in memory (estimated size 19.4 KB, free 465.4 KB)
```

As we know, the wholeTextFiles() function reads the file as a key/value pair, in which the file name is the key, and the content is the value. Let's get the file name:

```
>>> playData.keys().collect()
```

Here is the output:

```
[u'file:/home/pysparkbook/pysparkBookData/shakespearePlays.txt']
```

And now let's fetch the content of the file:

```
>>> playData.values().collect()
```

Here is the output:

```
[u"Love's Labour's Lost",
 u"A Midsummer Night's Dream",
 u'Much Ado About Nothing',
 u'As You Like It']
```

Step 6-1-3. Counting the Number of Lines in a File

We can count the number of lines in our file by using the count() function:

```
>>> playData.count()
```

Here is the output:

```
4
```

Step 6-1-4. Counting the Number of Characters on Each Line

To calculate the total number of characters in our file, we can calculate the number of characters in each line and then sum them. To calculate the total number of characters in each line, we can use the len() function. The len() function is defined in Python, and it calculates the number of elements in a sequence. For Python strings, the len() function will return the total number of characters in that string. Let's observe the workings of the len() function in the following example:

```
>>> pythonString = "My python"
>>> len(pythonString)
```

Here is the output:

```
9
```

We have created a string named pythonString. Then we've assigned a value, My python, to pythonString. The string My python has nine characters, including the space. Therefore, the len() function on the pythonString variable has returned 9 as the output. Now let's use the len() function in the RDD map() function:

```
>>> playDataLineLength = playData.map(lambda x : len(x))
```

```
>>> playDataLineLength.collect()
```

Here is the output:

```
[21, 25, 22, 14]
```

Finally, we have calculated the number of characters in each line. If we apply sum() on the playDataLineLength RDD, we will get the total number of characters in the file:

```
>>> totalNumberOfCharacters = playDataLineLength.sum()
>>> totalNumberOfCharacters
```

Here is the output:

```
82
```

Finally, we have calculated that the total number of characters in our file is 82. Remember, the count doesn't include the newline character.

Recipe 6-2. Write an RDD to a Simple Text File
Problem

You want to write an RDD to a simple text file.

Solution

In Recipe 6-1, you calculated the number of characters in each line as the RDD playDataLineLength. Now you want to save it in a text file.

We can save an RDD as a text file by using the saveAsTextFile() function. This method is defined on the RDD—not on SparkContext, as we saw in the case of the textFile() and wholeTextFiles() functions. You have to provide the output directory. The file name is not required. The directory name you are providing must not already exist; otherwise, the write operation will fail. The RDD exists in partitions. So PySpark will start many processes in parallel to write the file.

The saveAsTextFile() function takes two inputs. The first input is path, which is basically the path of the directory where the RDD has to be saved. The second argument is compressionCodecClass, an optional argument with a default value of None. We can use compression codecs such as Gzip to compress files and thereby provide more-efficient computations.

How It Works

So first let's start with the code for counting the number of characters in each line. We have already done it, but for the sake of clarity, I have provided the code for calculating the number of characters in each line again.

Step 6-2-1. Counting the Number of Characters on Each Line

Let's read the file and count the number of characters per line:

```
>>> playData = sc.textFile('/home/pysparkbook/pysparkBookData/
shakespearePlays.txt',4)
>>> playDataLineLength = playData.map(lambda x : len(x))
>>> playDataLineLength.collect()
```

Here is the output:

```
[21, 25, 22, 14]
```

Our playDataLineLength RDD has as its elements the number of characters in each line.

Step 6-2-2. Saving the RDD to a File

Now that we have the counted RDD, we want to save that RDD into a directory called savedData:

```
>>> playDataLineLength.saveAsTextFile('/home/pysparkbook/savedData')
```

Let's investigate the savedData directory. We will find that, inside the directory, there are five files. Four files—part-00000, part-00000, part-00000, and part-00000—will have the data, and the _SUCCESS file denotes that the data has been written successfully. Have you wondered why the data is in four files? The answer is that the playDataLineLength RDD has been distributed in four parts. Four parallel processes have been used to write the file.

Let's see what is inside those files. We will find that each data point is inside a separate file:

```
savedData$ cat part-00000
```

Here is the output:

```
21
```

```
savedData$ cat part-00001
```

Here is the output:

```
25
```

```
savedData$ cat part-00002
```

Here is the output:

```
22
```

```
savedData$ cat part-00003
```

Here is the output:

```
14
```

Recipe 6-3. Read a Directory
Problem

You want to read a directory.

Solution

In a directory, there are many files. You want to read the directory (all files at once).

Reading many files together from a directory is a very common task nowadays. To read a directory, we use the textFile() function or the wholetextFiles() function. The textFile() function reads small files in the directory and merges them. In contrast, the wholeTextFiles() function reads files as a key/value pair, with the name of file as the key, and the content of the file as the value.

You are provided with the directory name manyFiles. This directory consists of two files named playData1.txt and playData2.txt. Let's investigate the content of these files one by one:

```
manyFiles$ cat playData1.txt
```

Here is the output:

```
Love's Labour's Lost
A Midsummer Night's Dream
```

```
manyFiles$ cat playData2.txt
```

Here is the output:

```
Much Ado About Nothing
As You Like It
```

You job is to read all these files from the directory in one go.

How It Works

We will use both functions, textFile() and wholeTextFiles(), one at a time, to read the directory noted previously. Let's start with textFile().

Step 6-3-1. Reading a Directory by Using textFile()

In previous recipes, we provided the absolute file path as the input to the textFile() function in order to read the file. The best part of the textFile() function is that just by changing the path input, we can change how this function reads data. In order to read all files from a directory, we have to provide the absolute path of the directory as input to textFile(). The following line of code reads all the files in the manyFiles directory by using the textFile() function:

```
>>> manyFilePlayData = sc.textFile('/home/pysparkbook/pysparkBookData/
manyFiles',4)

>>> manyFilePlayData.collect()
```

Here is the output:

```
[u'Much Ado About Nothing',
 u'As You Like It',
 u"Love's Labour's Lost",
 u"A Midsummer Night's Dream"]
```

The output is very clear. The textFile() function has read all the files in the directory and merged the content in the files. It has created an RDD of the merged data.

Step 6-3-2. Reading a Directory by Using wholeTextFiles()

Let's now read the same set of files by using the wholeTextFiles() function. As we did with textFile(), here we also provide the path of the directory as one of the inputs to the wholeTextFiles() function. The following code shows the use of the wholeTextFiles() function to read a directory:

```
>>> manyFilePlayDataKeyValue = sc.wholeTextFiles('/home/pysparkbook/
pysparkBookData/manyFiles',4)
>>> manyFilePlayDataKeyValue.collect()
```

Here is the output:

```
[(u'file:/home/pysparkbook/pysparkBookData/manyFiles/playData2.txt', u'Much
Ado About Nothing\nAs You Like It\n'),
 (u'file:/home/pysparkbook/pysparkBookData/manyFiles/playData1.txt',
u"Love's Labour's Lost\nA Midsummer Night's Dream\n")]
```

The output has key/value pairs of file names and contents. Further, we have to process as required by the problem requirements.

Recipe 6-4. Read Data from HDFS

Problem

You want to read a file from HDFS by using PySpark.

Solution

HDFS is very good for storing high-volume files. You are given the file filamentData.csv in HDFS. This file is under the bookData directory. Our bookData is under the root directory of HDFS. You want to read this file by using PySpark.

To read a file from HDFS, we first need to know the fs.default.name property from the core-site.xml property file. We are going to get the core-site.xml file inside the Hadoop configuration directory. For us, the value of fs.default.name is hdfs://localhost:9746. The full path of our file in HDFS will be hdfs://localhost:9746/bookData/ filamentData.csv.

We can use the textFile() function to read the file from HDFS by using the full path of the file.

How It Works

Reading a file from HDFS is as easy as reading data from a local machine. In the following code line, we use the textFile() function to read the required file:

```
>>> filamentdata = sc.textFile('hdfs://localhost:9746/bookData/filamentData.csv',4)

>>> filamentdata.take(4)
```

Here is the output:

```
[u'filamentA,100W,605',
 u'filamentB,100W,683',
 u'filamentB,100W,691',
 u'filamentB,200W,561']
```

The result confirms that we are able to read our file from HDFS.

Recipe 6-5. Save RDD Data to HDFS

Problem

You want to save RDD data to HDFS.

Solution

RDD data can be saved to HDFS by using the saveAsTextFile() function. In Recipe 6-2, we saved the RDD in a file on the local file system. We are going to save the same RDD, playDataLineLength, to HDFS.

Similar to the way we worked with the textFile() function, we have to provide the full path of the file, including the NameNode URI, to saveAsTextFile() to write an RDD to HDFS.

How It Works

Because you are a keen reader and might not be looking for distractions, we'll start with writing code that counts the total number of characters in each line.

Step 6-5-1. Counting the Number of Characters on Each Line

```
>>> playData = sc.textFile('/home/muser/bData/shakespearePlays.txt',4)
>>> playDataLineLength = playData.map(lambda x : len(x))
>>> playDataLineLength.collect()
```

Here is the output:

```
[21, 25, 22, 14]
```

Step 6-5-2. Saving an RDD to HDFS

The playDataLineLength RDD is written using following code line:

```
>>> playDataLineLength.saveAsTextFile('hdfs://localhost:9746/savedData/')
```

We have saved the RDD in the savedData directory, which is inside the root directory of HDFS. Remember that the savedData directory didn't exist before we saved the data; otherwise, the saveAsTextFile() function would throw an error. We have saved the RDD. Now we are going to investigate the saved data. We will find five files in the savedData directory: part-00000, part-00001, part-0002, part-00003, and our _SUCCESS file.

Let's see the data of each file, one by one, by using the HDFS cat command. This command displays file data to the console:

```
$ hadoop fs -cat /savedData/part-00000
```

Here is the output:

```
21
```

```
$ hadoop fs -cat /savedData/part-00001
```

Here is the output:

```
25
```

```
$ hadoop fs -cat /savedData/part-00002
```

Here is the output:

```
22
```

```
$ hadoop fs -cat /savedData/part-00003
```

Here is the output:

```
14
```

Each file has a single data point because our RDD has four partitions.

Recipe 6-6. Read Data from a Sequential File
Problem

You want to read data from a sequential file.

Solution

A sequential file uses the key/value file format. Here, the key values are in binary format. This is a commonly used file format for Hadoop. The keys and values are types of the Hadoop Writable class.

We have data in a sequential file in HDFS, in the sequenceFileToRead directory inside the root directory. In the file inside the directory, we have the data in Table 6-1.

Table 6-1. *Sequential File Data*

Key	Value
p	20
q	30
r	20
m	25

We can read the sequence file by using the `sequenceFile()` method defined in the `SparkContext` class.

How It Works

The `sequenceFile()` function takes many arguments. Let me discuss some of them. The first argument is `path`, which is the path of the sequential file. The second argument is `keyClass`, which indicates the key class of data in the sequence file. The argument `valueClass` represents the data type of the values. Remember that the key and value classes are children of the Hadoop `Writable` classes.

```
>>> simpleRDD = sc.sequenceFile('hdfs://localhost:9746/sequenceFileToRead')
>>> simpleRDD.collect()
```

Here is the output:

```
[(u'p', 20),
 (u'q', 30),
 (u'r', 20),
 (u'm', 25)]
```

Finally, we have read our sequential file successfully.

Recipe 6-7. Write Data to a Sequential File
Problem

You want to write data into a sequential file.

Solution

Many times we like to save the results from PySpark processing to a sequence file. We have an RDD of subject data, as shown in Table 6-2, and you want to write it to a sequence file.

Table 6-2. *Student Subjects Data*

Student ID	Subjects
si1	Python
si3	Java
si1	Java
si2	Python
si3	Ruby
si4	C++
si5	C
si4	Python
si2	Java

How It Works

In this recipe, we are first going to create an RDD and then save it into a sequence file.

Step 6-7-1. Creating a Paired RDD

First, we'll create a list of tuples:

```
>>> subjectsData = [('si1','Python'),
...                 ('si3','Java'),
...                 ('si1','Java'),
...                 ('si2','Python'),
...                 ('si3','Ruby'),
...                 ('si4','C++'),
...                 ('si5','C'),
...                 ('si4','Python'),
...                 ('si2','Java')]
```

Next, we parallelize the data:

```
>>> subjectsPairedRDD = sc.parallelize(subjectsData, 4)

>>> subjectsPairedRDD.take(4)

[('si1', 'Python'),
 ('si3', 'Java'),
 ('si1', 'Java'),
 ('si2', 'Python'),
]
```

Step 6-7-2. Saving the RDD as a Sequence File

Finally, we write our paired RDD into a sequence file:

```
>>>subjectsPairedRDD.saveAsSequenceFile('hdfs://localhost:9746/sequenceFiles')
```

Our sequence file is placed in HDFS, inside the sequenceFiles directory. Now let's investigate the saved files:

```
$ hadoop fs -ls /sequenceFiles
```

We find five items:

```
-rw-r--r--  3 pysparkbook  supergroup 0 2017-05-22 00:18 /sequenceFiles /_SUCCESS
-rw-r--r--  3 pysparkbook supergroup        114 2017-05-22 00:18 /
sequenceFiles/part-00000
-rw-r--r--  3 pysparkbook supergroup        114 2017-05-22 00:18 /
sequenceFiles/part-00001
-rw-r--r--  3 pysparkbook supergroup        111 2017-05-22 00:18 /
sequenceFiles/part-00002
-rw-r--r--  3 pysparkbook supergroup        128 2017-05-22 00:18 /
sequenceFiles/part-00003
```

Here we can see that the files have been saved in four parts.

Recipe 6-8. Read a CSV File
Problem

You want to read a CSV file.

Solution

As we have mentioned, *CSV* stands for *comma-separated values*. In a CSV file, each line contains fields, which are separated by a delimiter. Generally, a comma (,) is used as the separating delimiter. But a delimiter could be a different character too.

We have seen that if PySpark reads a file, it creates an RDD of lines. So the best way to read a CSV file is to read it by using the textFile() function and then to parse each line by using a CSV parser in Python.

In Python, we can use the csv module to parse CSV lines. This module provides all the functionality for handling CSV data. We also use the reader() function, which returns a reader object. The reader object iterates over lines.

You have been given a CSV file of filaments data, the same filament data we used in Chapter 5. The lines of the filament data file look like the following:

```
filamentA,100W,605
filamentB,100W,683
filamentB,100W,691
filamentB,200W,561
filamentA,200W,530
```

We can see that each line in the file is separated by a comma. And so parsing each line by a CSV parser will complete our job.

▪ **Note**   You can read more about reading a CSV file by using PySpark at https://stackoverflow.com/questions/28782940/load-csv-file-with-spark.

How It Works

We will start with writing a Python function that will parse each line.

Step 6-8-1. Writing a Python Function to Parse CSV Lines

We are going to write a function named parseCSV(). This function will take each line and parse it to return a list:

```
>>> import csv
>>> import  StringIO
>>> def parseCSV(csvRow) :
...      data = StringIO.StringIO(csvRow)
...      dataReader = csv.reader(data)
...      return(dataReader.next())
```

Our function takes one line at a time. Each line is taken by the StringIO function defined in the StringIO module. Using the StringIO module, we can create a file-like object. We pass the data object to the reader() function of the csv module. It will return a reader object. The next() function will return a list of fields that are separated by commas. Let's check whether our function is working correctly:

```
>>> csvRow = "p,s,r,p"
>>> parseCSV(csvRow)
```

Here is the output:

```
['p', 's', 'r', 'p']
```

We have created a string, csvRow. We can see that the fields in our string are separated by commas. Now we are parsing that line by using our function and getting a list as a result. So it is confirmed that our function meets our expectations.

Step 6-8-2. Creating a Paired RDD

We will read the file and parse its lines. The following code will read the file, parse the lines, and return the required data:

```
>>> filamentRDD =sc.textFile('/home/pysparkbook/pysparkBookData
filamentData.csv',4)
>>> filamentRDDCSV = filamentRDD.map(parseCSV)

>>> filamentRDDCSV.take(4)
```

Here is the output:

```
[['filamentA', '100W', '605'],
 ['filamentB', '100W', '683'],
 ['filamentB', '100W', '691'],
 ['filamentB', '200W', '561']]
```

The final output returns a nested list. Each inside list consists of parsed lines.

Recipe 6-9. Write an RDD to a CSV File

Problem

You want to write an RDD to a CSV file.

Solution

Writing data to a CSV file requires that we transform the list of fields into strings of comma-separated fields. A list can be transformed to a string as elements are concatenated.

How It Works
Step 6-9-1. Creating a Function to Convert a List into a String

Let's create a function named createCSV, which will take a list and return a string by joining the elements of the list that are separated by commas.

```
>>> import csv
>>> import StringIO
>>> def createCSV(dataList):
...        data = StringIO.StringIO()
...        dataWriter = csv.writer(data,lineterminator='')
...        dataWriter.writerow(dataList)
...        return (data.getvalue())
```

The StringIO() function returns a file-like object. Then the writerow() function of the csv module transforms it into a string. Let's observe the action of the createCSV() function:

```
>>> listData = ['p','q','r','s']

>>> createCSV(listData)
```

Here is the output:

```
'p,q,r,s'
```

The listData list has four elements. We provide listData as input to the createCSV() function. And, finally, we get a string, which is created by concatenating elements of the list separated by commas.

Step 6-9-2. Saving Data to a File

Our problem asks us to save the data to a file. But let's create an RDD and then save the data. We are going to create an RDD from our simpleData nested list:

```
>>> simpleData = [['p',20],
...                ['q',30],
...                ['r',20],
...                ['m',25]]

>>> simpleRDD = sc.parallelize(simpleData,4)
>>> simpleRDD.take(4)
```

Here is the output:

```
[['p', 20],
 ['q', 30],
 ['r', 20],
 ['m', 25]]
```

We have created our required RDD. Now, using the map() function, we'll transform our data into the required format for saving:

```
simpleRDDLines = simpleRDD.map( createCSV)
simpleRDDLines.take(4)
simpleRDDLines.saveAsTextFile('/home/pysparkbook/csvData/')
```

We have saved our data as a CSV file in the csvData directory.

Recipe 6-10. Read a JSON File

Problem

You want to read a JSON file by using PySpark.

Solution

As noted previously in the chapter, *JSON* stands for *JavaScript Object Notation*. The popularity of JSON data files has increased over the decades. It is a data interchange format. Nearly all programming languages support reading and writing JSON. JSON format has like fields name and values pair. Two fields name and value pair is separated by comma (,), and each field name and associated value are separated by a colon (:). Data scientists often get data from clients in JSON format.

You have been given a JSON file, tempData.json. JSON files have a .json extension. This file has data in the following format:

```
$ cat tempData.json
```

Here is the output:

```
{"Time":"6AM",  "Temperature":15}
{"Time":"8AM",  "Temperature":16}
{"Time":"10AM", "Temperature":17}
{"Time":"12AM", "Temperature":17}
{"Time":"2PM",  "Temperature":18}
{"Time":"4PM",  "Temperature":17}
{"Time":"6PM",  "Temperature":16}
{"Time":"8PM",  "Temperature":14}
```

Our JSON file consists of two keys, Time and Temperature, and their associated values. We can see that for a particular field, the field name and value are associated by a colon. And the two fields' name/value pairs are separated by a comma.

What is the big deal about reading a JSON file in PySpark? Now you understand that whenever we read a JSON file, it will create an RDD of lines and then parse the data line by line. So we need a parser to parse the lines, and more precisely, we need a JSON data parser.

How It Works

Many Python modules are available for JSON parsing, but we are going to use the json Python module. This module provides many utilities for working with JSON format.

Step 6-10-1. Creating a Function to Parse JSON Data

Let's create a function that will parse JSON data and return a Python list for each line:

```
>>> import json

>>> def jsonParse(dataLine):
...        parsedDict = json.loads(dataLine)
...        valueData = parsedDict.values()
...        return(valueData)
```

The jsonParse() function will take a string in JSON format and return a list. Input to the jsonParse() function is passed to the json.loads() function. The json.loads() function takes a JSON string as an argument and returns a dictionary. Then we fetch values in the next line and finally return it. We must test our function:

```
>>> jsonData = '{"Time":"6AM",  "Temperature":15}'
```

We create a JSON string, jsonData:

```
>>> jsonParsedData = jsonParse(jsonData)
>>> print jsonParsedData
```

Here is the output:

```
[15, '6AM']
```

To parse our JSON string, we use the jsonParse() function and get the Python list jsonParsedData.

Step 6-10-2. Reading the File

We are going to read the file by using textFile():

```
>>> tempData = sc.textFile("/home/pysparkbook//pysparkBookData/tempData.json",4)

>>> tempData.take(4)
```

Here is the output:

```
[u'{"Time":"6AM",    "Temperature":15}',
 u'{"Time":"8AM",    "Temperature":16}',
 u'{"Time":"10AM",   "Temperature":17}',
 u'{"Time":"12AM",   "Temperature":17}']
```

We have read the file, and the output tells us that the RDD is a line of JSON strings.

Step 6-10-3. Creating a Paired RDD

Now we have to parse each line. This can be done by passing the jsonParse() function as an argument to the map() function on our RDD tempData:

```
>>> tempDataParsed = tempData.map(jsonParse)
>>> tempDataParsed.take(4)
```

Here is the output:

```
[[15, u'6AM'],
 [16, u'8AM'],
 [17, u'10AM'],
 [17, u'12AM']]
```

Finally, we have the required result.

Recipe 6-11. Write an RDD to a JSON File
Problem

You want to write an RDD to a JSON file.

Solution

JSON string objects can be created from a Python dictionary by using the json module.

How It Works

Step 6-11-1. Creating a Function That Takes a List and Returns a JSON String

We are going to create a function, createJSON(), that will take a Python list and return a JSON string:

```
>>> def createJSON(data):
...     dataDict = {}
...     dataDict['Name'] = data[0]
...     dataDict['Age'] = data[1]
...     return(json.dumps(dataDict))
```

Inside the createJSON() function, we first create a dataDict dictionary. Then we pass that dictionary to the dumps() function of the json module. This function returns a JSON string object. The following line of code will test our function:

```
>>> nameAgeList = ['Arun',22]

>>> createJSON(nameAgeList)
```

Here is the output:

```
'{"Age": 22, "Name": "Arun"}'
```

Our function returns a JSON string object.

Step 6-11-2. Saving Data in JSON Format

We are going to follow the same procedure as before. First, we will create an RDD. Then we will write that RDD to a file in JSON format.

```
>>> nameAgeData = [['Arun',22],
...                          ['Bony',35],
...                          ['Juna',29]]
>>> nameAgeRDD = sc.parallelize(nameAgeData,3)

>>> nameAgeRDD.collect()
```

Here is the output:

```
[['Arun', 22],
 ['Bony', 35],
 ['Juna', 29]]
```

We have created an RDD; now we have to transform this data into JSON strings. The createJSON() function is being passed to the map() function as an argument. The map() function works on our nameAgeRDD RDD.

```
>>> nameAgeJSON = nameAgeRDD.map(createJSON)
>>> nameAgeJSON.collect()
```

Here is the output:

```
['{"Age": 22, "Name": "Arun"}',
 '{"Age": 35, "Name": "Bony"}',
 '{"Age": 29, "Name": "Juna"}']
```

The nameAgeJSON RDD elements are JSON strings. Now, using the saveAsTextFile() function, we save the nameAgeJSON RDD to the jsonDir directory.

```
>>> nameAgeJSON.saveAsTextFile('/home/pysparkbook/jsonDir/')
```

We are going to investigate the jsonDir directory. We use the ls command to find that there are four files:

```
jsonDir$ ls
```

Here is the output:

```
part-00000
part-00001
part-00002
_SUCCESS
```

```
jsonDir$ cat part-00000
```

Here is the output:

```
{"Age": 22, "Name": "Arun"}
```

The part-00000 file contains the first element of the RDD.

Recipe 6-12. Read Table Data from HBase by Using PySpark

Problem

You want to read table data from HBase.

Solution

We have been given a data table named pysparkTable in HBase. You want to read that table by using PySpark. The data in pysparkTable is shown in Table 6-3.

Table 6-3. *pysparkTable Data*

RowID	btcf1	btcf2
	btc1	btc2
00001	c11	c21
00002	c12	c22
00003	c13	c23
00004	c14	c24

Let's explain this table data. The pysparkTable table consists of four rows and two column families, btcf1 and btcf2. Column btc1 is under column family btcf1, and column btc2 is under column family btcf2. Remember that the code presented later in this section will work only with spark-1.6 and the older PySpark versions. Try tweaking the code to run on PySpark version 2.*x*.

We are going to use the newAPIHadoopRDD() function, which is defined on SparkContext sc. This function returns a paired RDD. Table 6-4 lists the arguments of the newAPIHadoopRDD() function.

Table 6-4. *Arguments of the newAPIHadoopRDD() Function*

Argument	Description
inputFormatClass	Fully qualified classname of Hadoop InputFormat (e.g., "org.apache.hadoop.mapreduce.lib.input.TextInputFormat")
keyClass	
valueClass	Fully qualified classname of value Writable class (e.g., "org.apache.hadoop.io.LongWritable")
keyConverter	Key converter
valueConverter	Value converter
conf	Hadoop configuration, passed in as a dict
batchSize	The number of Python objects represented as a single Java object. (default 0, choose batchSize automatically)

How It Works

Let's first define all the arguments that have to be passed into our
newAPIHadoopRDD() function:

```
>>> hostName = 'localhost'

>>> tableName = 'pysparkBookTable'

>>>    ourInputFormatClass='org.apache.hadoop.hbase.mapreduce.TableInputFormat'
>>> ourKeyClass='org.apache.hadoop.hbase.io.ImmutableBytesWritable'
>>> ourValueClass='org.apache.hadoop.hbase.client.Result'
>>> ourKeyConverter='org.apache.spark.examples.pythonconverters.
    ImmutableBytesWritableToStringConverter'
>>> ourValueConverter='org.apache.spark.examples.pythonconverters.
    HBaseResultToStringConverter'
>>> configuration = {}
>>> configuration['hbase.mapreduce.inputtable'] = tableName
>>> configuration['hbase.zookeeper.quorum'] = hostName
```

Now it is time to call the newAPIHadoopRDD() function with its arguments:

```
>>> tableRDDfromHBase = sc.newAPIHadoopRDD(
...                         inputFormatClass = ourInputFormatClass,
...                         keyClass = ourKeyClass,
...                         valueClass = ourValueClass,
```

```
...                          keyConverter = ourKeyConverter,
...                          valueConverter = ourValueConverter,
...                          conf = configuration
...                     )
```

Let's see how our paired RDD, tableRDDfromHBase, looks:

```
>>> tableRDDfromHBase.take(2)
```

Here is the output:

```
[(u'00001', u'{"qualifier" : "btc1", "timestamp" : "1496715394968",
"columnFamily" : "btcf1", "row" : "00001", "type" : "Put", "value" : "c11"}\
n{"qualifier" : "btc2", "timestamp" : "1496715408865", "columnFamily" :
"btcf2", "row" : "00001", "type" : "Put", "value" : "c21"}'), (u'00002',
u'{"qualifier" : "btc1", "timestamp" : "1496715423206", "columnFamily" :
"btcf1", "row" : "00002", "type" : "Put", "value" : "c12"}\n{"qualifier" :
"btc2", "timestamp" : "1496715436087", "columnFamily" : "btcf2", "row" :
"00002", "type" : "Put", "value" : "c22"}')]
```

The paired RDD tableRDDfromHBase has RowID as a key. The columns and column classifiers are JSON strings, which is the value part. In a previous recipe, we solved the problem of reading JSON files.

■ **Note**   Remember, Recipe 6-12 code will work with only spark-1.6 and before. You can get the code on GitHub at https://github.com/apache/spark/blob/ ed9d80385486cd39a84a689ef467795262af919a/examples/src/main/python/hbase_ inputformat.py.

There is another twist. We are using many classes, so we have to add some JARs while starting the PySpark shell. The following are the JAR files:

- spark-examples-1.6.0-hadoop2.6.0.jar

- hbase-client-1.2.4.jar

- hbase-common-1.2.4.jar

The following is the command to start the PySpark shell:

```
pyspark --jars 'spark-examples-1.6.0-hadoop2.6.0.jar','/hbase-client-
1.2.4.jar','hbase-common-1.2.4.jar'
```

CHAPTER 7

■ ■ ■

Optimizing PySpark and PySpark Streaming

Spark is a distributed framework for facilitating parallel processing. The parallel algorithms require computation and communication between machines. While communicating, machines send or exchange data. This is also known as *shuffling*.

Writing code is easy. But writing a program that is efficient and easy to understand by others requires more effort. This chapter presents some techniques for making the PySpark program clearer and more efficient.

Making decisions is a day-to-day activity. Our data-conscious population wants to include data analysis and result inference at the time of decision-making. We can gather data and do analysis, and we have done all of that in previous chapters. But people are becoming more interested in analyzing data as it is coming in. This means people are becoming more interested in analyzing *streaming data*.

Handling streaming data requires more robust systems and proper algorithms. The fault-tolerance of batch-processing systems is sometimes less complex than the fault-tolerance of a streaming-execution system. This is because in stream data processing, we are reading data from an outer source, running execution, and saving the results, all at the same time. More activities translate into a greater chance of failure.

In PySpark, streaming data is handled by its library, *PySpark Streaming*. PySpark Streaming is a set of APIs that provide a wrapper over PySpark Core. These APIs are efficient and deal with many aspects of fault-tolerance too. We are going to read streaming data from the console by using PySpark and then analyze it. We are also going to read data from Apache Kafka by using PySpark Streaming and then analyze the data.

This chapter covers the following recipes:

> Recipe 7-1. Optimize the page-rank algorithm using PySpark code
>
> Recipe 7-2. Implement the k-nearest neighbors algorithm using PySpark
>
> Recipe 7-3. Read streaming data from the console using PySpark Streaming
>
> Recipe 7-4. Integrate Apache Kafka with PySpark Streaming, and read and analyze the data

© Raju Kumar Mishra 2018
R. K. Mishra, *PySpark Recipes*, https://doi.org/10.1007/978-1-4842-3141-8_7

Recipe 7-1. Optimize the Page-Rank Algorithm by Using PySpark Code

Problem

You want to optimize the page-rank algorithm by using the PySpark code you wrote in Chapter 5.

Solution

We already implemented the page-rank algorithm in Chapter 5. Can we optimize the code? Whenever we start looking at every line of code, we can try to optimize it. In this recipe, we are going to optimize RDD joining. But you might want to look at other parts of the code and try to optimize different lines of the program.

Paired RDD joining is one of the costliest activities in PySpark or any distributed framework. Why? Because for any key in the first RDD, the system looks for all the keys in different data partitions of other RDDs. And then after lookup, the data is shuffled. If we repartition the data in such a way that the similar keys come to the same machine, then the data shuffle will be reduced. PySpark provides partitioners for the same purpose. We can use the partitionBy() function with the partitioner of our choice, so that similar keys occur on the same machine and less data shuffles. This results in improved speed of code execution.

How It Works

We already discussed the page-rank algorithm in detail in Chapter 5; if you need a refresher, please review that detailed discussion. In this recipe, I discuss only the code lines that have been modified. The modified code lines are in bold font. Let's look at those lines:

```
>>> pageLinks = [['a' ,['b','c','d']],
...              ['b', ['d','c']],
...              ['c', ['b']],
...              ['d', ['a','c']]]
>>> pageRanks = [['a',1],
...              ['b',1],
...              ['c',1],
...              ['d',1]]

>>> numIter = 20
```

```
>>> pageRanksRDD  = sc.parallelize(pageRanks, 2).partitionBy(2,hash).persist()
>>> pageLinksRDD  = sc.parallelize(pageLinks, 2).partitionBy(2,hash).persist()
```

In the bold code lines, I have added the partitionBy() method. The first argument of partitionBy() is 2, which tells us that the data is split into two partitions. There is another argument value that we have provided, hash. This argument value is the partitioner name. We repartition the data into two partitions, using the hash partitioner. The hash partitioner uses a hashing technique to repartition the data. We are using the same technique on both pageRanksRDD and pageLinksRDD, so the same key will go to the same machine. Therefore, we are tackling the shuffling problem. The other parts of the code are the same as in Chapter 5.

```
>>> s = 0.85

>>> def rankContribution(uris, rank):
...     numberOfUris = len(uris)
...     rankContribution = float(rank) / numberOfUris
...     newrank =[]
...     for uri in uris:
...             newrank.append((uri, rankContribution))
...     return newrank

>>> for i in range(numIter):
...         linksRank = pageLinksRDD.join(pageRanksRDD)
...     contributedRDD = linksRank.flatMap(lambda x : rankContribution(x[1]
[0],x[1][1]))
...     sumRanks = contributedRDD.reduceByKey(lambda v1,v2 : v1+v2)
...     pageRanksRDD = sumRanks.map(lambda x : (x[0],(1-s)+s*x[1]))
...
>>> pageRanksRDD.collect()
```

Here is the output:

```
[('b', 1.357243795127982),
 ('d', 0.8746512999550939),
 ('a', 0.5217268024809147),
 ('c', 1.2463781024360086)]
```

We have the final answer for the page-rank implementation.

Recipe 7-2. Implement the k-Nearest Neighbors Algorithm by Using PySpark

Problem

You want to implement the k-nearest neighbors (KNN) algorithm by using PySpark.

Solution

The *k-nearest neighbors* algorithm is one of the simplest data-classification algorithms. The similarity between two data points is measured on the basis of the distance between two points.

We have been given a dataset of nine records. This dataset is shown in Table 7-1. In this table, you can see a column named RN. That column indicates the *record number*. This is not part of the data; the record number is given to help you understand the KNN algorithm.

Table 7-1. *Data for Classification by KNN*

RN	ivs1	ivs2	ivs3	Group
1	3.09	1.97	3.73	group1
2	2.96	2.15	4.16	group1
3	2.87	1.93	4.39	group1
4	3.02	1.55	4.43	group1
5	1.80	3.65	2.08	group2
6	1.36	4.43	1.95	group2
7	1.71	4.35	1.94	group2
8	1.03	3.75	2.12	group2
9	2.30	3.59	1.99	group2

Let's say that we have a record (ivs1 = 2.5, ivs2 = 1.7, ivs3 = 4.2). We will call this record *new record*. We have to classify this record; it will be in either group1 or group2.

To classify the record, we'll use the KNN algorithm. Here are the steps:

1. Decide the k.

   k is the number of nearest neighbors we are going to choose for deciding the class of the new record. Let's say k is 5.

2. Find the distance of the new record from each record in the data. The distance calculation is done using the Euclidean distance method, as shown in Table 7-2.

Table 7-2. *Distance Calculation*

RN	Distance Calculation with New Record	Distance
1	sqrt((3.09-2.5)^2 + (1.97 -1.7)^2 + (3.73 -4.2) ^2)	0.80
2	sqrt((2.96-2.5)^2 + (2.15 -1.7)^2 + (4.16 -4.2) ^2)	0.65
3	sqrt((2.87-2.5)^2 + (1.93 -1.7)^2 + (4.39 -4.2) ^2)	0.47
4	sqrt((3.02-2.5)^2 + (1.55 -1.7)^2 + (4.43 -4.2) ^2)	0.58
5	sqrt((1.80-2.5)^2 + (3.65 -1.7)^2 + (2.08 -4.2) ^2)	2.96
6	sqrt((1.36-2.5)^2 + (4.43 -1.7)^2 + (1.95 -4.2) ^2)	3.71
7	sqrt((1.71-2.5)^2 + (4.35 -1.7)^2 + (1.94 -4.2) ^2)	3.57
8	sqrt((1.03-2.5)^2 + (3.75 -1.7)^2 + (2.12 -4.2) ^2)	3.26
9	sqrt((2.30-2.5)^2 + (3.59 -1.7)^2 + (1.99 -4.2) ^2)	2.91

In this table, we have calculated the distance between the new record and other records. The third column is the distance. The distance value of the first row of this table is the distance between the new record and record 1.

3. Sort the distances.

Sorting is required now. And we have to sort the distances in increasing order. Simultaneously, we have to maintain the association between the RN column and the Distance column. In Table 7-3, we have sorted the Distance column, which is still associated with the RN and Group columns.

Table 7-3. *Distance Calculation*

RN	Distance		RN	Distance	Group
3	0.47		3	0.47	group1
4	0.58		4	0.58	group1
2	0.65		2	0.65	group1
1	0.8		1	0.8	group1
9	2.91		9	2.91	group2
5	2.96		5	2.96	group2
8	3.26		8	3.26	group2
7	3.57		7	3.57	group2
6	3.71		6	3.71	group2

4.   Find the k-nearest neighbors.

Now that we have sorted the Distance column, we have to identify the neighbors of the new record. What do I mean by *neighbors* here? *Neighbors* are those records in the table that are near the new record. *Near* means having less distance between two nodes. Now look for the five nearest neighbors in Table 7-3. For the new record, records 3, 4, 2, 1, and 9 are neighbors. The group for records 3, 4, 2, and 1 is group1. The group for record 9 is group2. The majority of neighbors are from group1. Therefore, we can classify the new record in group1.

We have discussed the KNN algorithm in detail. Let's see how to implement it by using PySpark. We are going to implement KNN in a naive way and then we will optimize it in the "How It Works" section.

First, we are going to calculate the distance between two tuples. We'll write a Python function, distanceBetweenTuples(). This function will take two tuples, calculate the distance between them, and return that distance:

```
>>> def distanceBetweenTuples(data1 , data2) :
...     squaredSum = 0.0
...     for i in range(len(data1)):
...         squaredSum = squaredSum + (data1[i] - data2[i])**2
...     return(squaredSum**0.5)
```

Now that we've written the function to calculate the distance, let's test it:

```
>>> pythonTuple1 = (1.2, 3.4, 3.2)
>>> pythonTuple2 = (2.4, 2.2, 4.2)
>>> distanceBetweenTuples(pythonTuple1, pythonTuple2)
```

Here is the output:

```
1.9697715603592207
```

Our method has been tested. It is a general function. We can run it for tuples of length 4 or 5 also. In the following lines of code, we'll create a list. The elements of this list are tuples. Each tuple has two elements. The first element is itself a tuple of data. The second element of each tuple is the group associated with each tuple.

```
>>> knnDataList = [((3.09,1.97,3.73),'group1'),
...                ((2.96,2.15,4.16),'group1'),
...                ((2.87,1.93,4.39),'group1'),
...                ((3.02,1.55,4.43),'group1'),
...                ((1.80,3.65,2.08),'group2'),
...                ((1.36,4.43,1.95),'group2'),
```

```
...                          ((1.71,4.35,1.94),'group2'),
...                          ((1.03,3.75,2.12),'group2'),
...                          ((2.30,3.59,1.99),'group2')]

>>> knnDataRDD = sc.parallelize(knnDataList, 4)
```

The data has been parallelized. We define newRecord as [(2.5, 1.7, 4.2)]:

```
>>> newRecord = [(2.5, 1.7, 4.2)]
>>> newRecordRDD = sc.parallelize(newRecord, 1)

>>> cartesianDataRDD = knnDataRDD.cartesian(newRecordRDD)
>>> cartesianDataRDD.take(5)
```

Here is the output:

```
[(((3.09, 1.97, 3.73), 'group1'), (2.5, 1.7, 4.2)),
 (((2.96, 2.15, 4.16), 'group1'), (2.5, 1.7, 4.2)),
 (((2.87, 1.93, 4.39), 'group1'), (2.5, 1.7, 4.2)),
 (((3.02, 1.55, 4.43), 'group1'), (2.5, 1.7, 4.2)),
 (((1.8, 3.65, 2.08), 'group2'), (2.5, 1.7, 4.2))]
```

We have created a Cartesian by using the older record data and the new record data. You might be wondering why I have created this Cartesian. at the time of defining the list knnDataList. In a real case, you would have a large file. That file might be distributed also. So for that condition, we'd have to read the file first and then create the Cartesian. After creating the Cartesian, we have the older data and the new record data in the same row, so we can easily calculate the distance with the map() method:

```
>>> K = 5

>>> groupAndDistanceRDD = cartesianDataRDD.map(lambda data : (data[0][1]
,distanceBetweenTuples(data[0][0], data[1])))

>>> groupAndDistanceRDD.take(5)
```

Here is the output:

```
[('group1', 0.8011866199581719),
 ('group1', 0.6447480127925947),
 ('group1', 0.47528938553264566),
 ('group1', 0.5880476171195661),
 ('group2', 2.9642705679475347)]
```

We have calculated the RDD groupAndDistanceRDD; its first element is the group, and the second element is the distance between the new record and older records. We have to sort it now in increasing order of distance. You might remember the takeOrdered() function described in Chapter 4. So let's get five groups in increasing order of distance:

```
>>> ourClasses = groupAndDistanceRDD.takeOrdered(K, key = lambda data : data[1])
>>> ourClasses
```

Here is the output:

```
[('group1', 0.47528938553264566),
 ('group1', 0.5880476171195661),
 ('group1', 0.6447480127925947),
 ('group1', 0.8011866199581719),
 ('group2', 2.9148241799463652)]
```

Using the takeOrdered() method, we have fetched five elements of the RDD, with the distance in increasing order. We have to find the group that is in the majority. So we have to first fetch only the group part and then we have to find the most frequent group:

```
>>> ourClassesGroup = [data[0] for data in ourClasses]
>>> ourClassesGroup
```

Here is the output:

```
['group1', 'group1', 'group1', 'group1', 'group2']
```

The group part has been fetched. The most frequent group can be found using the max() Python function as follows:

```
>>> max(ourClassesGroup,key=ourClassesGroup.count)
```

Here is the output:

```
'group1'
```

We finally have the group of the new record, and that is group1.

You might be thinking that now that we have implemented KNN, what's next? Next, we should optimize the code. Let me say that again. We can optimize different aspects of this code. For this example, we'll use the broadcasting technique using the broadcast variable. This is a very good technique for optimizing code.

The Cartesian has been applied to join the older records with the new record. PySpark provides another way to achieve a similar result. We can send the new record to every executor before. This new record data will be available to each executor, and they can use it for distance calculations. We can send the new record tuple to all the executors as a broadcast variable.

Broadcast variables are shared and read-only variables. *Read-only* means executors cannot change the value of a broadcast variable; they can only read the value of it. In PySpark, we create a broadcast variable by using the broadcast() function. This broadcast() function is defined on SparkContext. We know that in the PySpark console, we have SparkContext as sc. We are going to reimplement the KNN by using the broadcast technique.

How It Works

We have already discussed most of the code. Therefore, I will keep the discussion short in the coming steps.

Step 7-2-1. Creating a Function to Calculate the Distance Between Two Tuples

```
>>> def distanceBetweenTuples(data1 , data2) :
...     squaredSum = 0.0
...     for i in range(len(data1)):
...         squaredSum = squaredSum + (data1[i] - data2[i])**2
...     return(squaredSum**0.5)
```

We have already created and tested this method.

Step 7-2-2. Creating a List of Given Records and Transforming It to an RDD

```
>>> knnDataList = [((3.09,1.97,3.73),'group1'),
...                ((2.96,2.15,4.16),'group1'),
...                ((2.87,1.93,4.39),'group1'),
...                ((3.02,1.55,4.43),'group1'),
...                ((1.80,3.65,2.08),'group2'),
...                ((1.36,4.43,1.95),'group2'),
...                ((1.71,4.35,1.94),'group2'),
...                ((1.03,3.75,2.12),'group2'),
...                ((2.30,3.59,1.99),'group2')]
>>> K = 5
```

We again want to go for five neighbors in order to determine the group of the new record. We also parallelize the data and transform it to an RDD of four partitions:

```
>>> knnDataRDD = sc.parallelize(knnDataList, 4)

>>> knnDataRDD.take(5)
```

Here is the output:

```
[((3.09, 1.97, 3.73), 'group1'),
 ((2.96, 2.15, 4.16), 'group1'),
 ((2.87, 1.93, 4.39), 'group1'),
 ((3.02, 1.55, 4.43), 'group1'),
 ((1.8, 3.65, 2.08), 'group2')]
```

Step 7-2-3. Broadcasting the Record Value

Now we have to create the required new record:

```
>>> newRecord = [(2.5, 1.7, 4.2)]
```

Broadcasting the new record will be done by the broadcast() method, which is defined on the SparkContext object. We can read the value of the broadcasted variable by using the value attribute:

```
>>> broadCastedValue = sc.broadcast(newRecord)
>>> broadCastedValue.value
```

Here is the output:

```
[(2.5, 1.7, 4.2)]
```

We can see that it returns the data as we have broadcasted it. But we need the tuple that we can get by fetching the first element of the list:

```
>>> broadCastedValue.value[0]
```

Here is the output:

```
(2.5, 1.7, 4.2)
```

Step 7-2-4. Broadcasting the Record Value

After broadcast, we have to create an RDD that will be an RDD of tuples. Each tuple's first element is the group, and the second element is the distance:

```
>>> groupAndDistanceRDD = knnDataRDD.map(lambda data : (data[1]
,distanceBetweenTuples(data[0], tuple(broadCastedValue.value[0]))))
```

To calculate the distance, we use the distanceBetweenTuples() method:

```
>>> groupAndDistanceRDD.take(5)
```

Here is the output:

```
[('group1', 0.8011866199581719),
 ('group1', 0.6447480127925947),
 ('group1', 0.47528938553264566),
 ('group1', 0.5880476171195661),
 ('group2', 2.9642705679475347)]
```

The requirement is achieved.

Step 7-2-5. Finding the Class of a New Record

We'll find the class of the new record in the same way as we did in its naive part:

```
>>> ourClasses = groupAndDistanceRDD.takeOrdered(K, key = lambda data :
data[1])
>>> ourClasses
```

Here is the output:

```
[('group1', 0.47528938553264566),
 ('group1', 0.5880476171195661),
 ('group1', 0.6447480127925947),
 ('group1', 0.8011866199581719),
 ('group2', 2.9148241799463652)]
```

```
>>> ourClassesGroup = [data[0] for data in ourClasses]
>>> ourClassesGroup
```

Here is the output:

```
['group1', 'group1', 'group1', 'group1', 'group2']
```

```
>>> max(ourClassesGroup,key=ourClassesGroup.count)
```

Here is the output:

```
'group1'
```

We can see that the class of the new record is group1.

> ■ **Note** You can read more about the PySpark broadcast variable in the following blogs:
>
> https://stackoverflow.com/questions/34499650/spark-broadcast-vs-join
>
> https://stackoverflow.com/questions/34864751/apache-spark-broadcast-variables-are-type-broadcast-not-a-rdd
>
> https://stackoverflow.com/questions/40685469/in-spark-how-does-broadcast-work/40694867
>
> You can find more on KNN on Wikipedia, at https://en.wikipedia.org/wiki/K-nearest_neighbors_algorithm.

Recipe 7-3. Read Streaming Data from the Console Using PySpark Streaming
Problem

You want to read streaming data from the console by using PySpark Streaming.

Solution

Netcat is network utility software. It can read and write data using TCP or UDP. It can be used as a client or server or both. Many options can be provided. We can provide listen mode by using the -l option. And the s option can be used to keep inbound sockets open for multiple connections. We are going to use a Netcat server to create a console source for data.

Open a terminal and start a Netcat server by using the command nc -lk 55342. Figure 7-1 depicts the starting of a Netcat server.

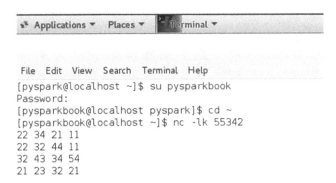

Figure 7-1. Netcat server in a console

174

PySpark Streaming can read data from many sources. In order to read data from a console source, we are going to use the socketTextStream() function. This function is defined on the StreamingContext object. Just as SparkContext is a way to enter PySpark, StreamingContext is a way to enter the PySpark Streaming library and use the APIs available in the package.

Just consider that a source produces numeric data. A source puts data on the console. A line of numeric data is created with time. We have to read the console data by using PySpark Streaming, do a summation of each line, and print the result on the console. The data is shown here:

```
22 34 21 11
22 32 44 11
32 43 34 54
21 23 32 21
```

The result looks like this:

```
88.0
109.0
163.0
97.0
```

How It Works
Step 7-3-1. Starting a Netcat Server

A Netcat server can be started by using the nc command. After starting the server, we can type some data on it. Or we can type some data after connecting PySpark Streaming to this server. Here is the data:

```
$ nc -lk  55342
22 34 21 11
22 32 44 11
32 43 34 54
21 23 32 21
22 32 21 32
21 23 32 21
22 32 21 32
22 32 21 32
32 44 54 32
```

Step 7-3-2. Defining a Function to Sum Row Data

In this step, we'll define a function named stringToNumberSum(). PySpark will read the data from the Netcat server as a string. Sometimes a string might have leading and trailing spaces. So first we have to remove the spaces, which can be done by using the strip() function defined on the String data type. Sometimes we might get a blank string; we can deal with that by using an if block. If we get a blank string, we return the None data type. But if a string is not empty, we split the data points by using the split() function defined on the Python string. But remember, even after splitting the data points, they are of the String type. We can change it to a float type by using the float() function or integers by using the int() function. A string is transformed to a float by using list comprehension. And finally, the sum() function is used to calculate the sum of a list of elements, and the final result is returned.

```
>>> def stringToNumberSum(data):
...         removedSpaceData = data.strip()
...         if   removedSpaceData == '' :
...             return(None)
...         splittedData =  removedSpaceData.split(' ')
...         numData =  [float(x) for x in splittedData]
...         sumOfData = sum(numData)
...         return (sumOfData)
```

Step 7-3-3. Reading Data from the Netcat Server and Calculating the Sum

We should start reading data from the Netcat server by using the PySpark Streaming library. In order to use the APIs defined in the Streaming library, we have to first import StreamingContext and create an object out of it:

```
>>> from pyspark.streaming import StreamingContext
>>> pysparkBookStreamingContext = StreamingContext(sc, 10)
```

Let me explain the preceding code. First, we import StreamingContext. Then we create an object of StreamingContext. In StreamingContex(), the first argument is the SparkContext object sc, and the second argument is the time in seconds. Using this StreamingContext object pysparkBookStreamingContext, we are going to read data from the Netcat server by using the socketTextStream() function:

```
>>> consoleStreamingData = pysparkBookStreamingContext.socketTextStream(
...                                         hostname = 'localhost',
...                                         port = 55342
...                                         )
```

The first argument of the socketTextStream() method is the hostname where the Netcat server is running. The second argument is the port of the server.

```
>>> sumedData = consoleStreamingData.map(stringToNumberSum)
>>> sumedData.pprint()
```

We do the summation using map(), and then we print it to the console by using the pprint() function:

```
>>> pysparkBookStreamingContext.start() ;pysparkBookStreamingContext.
awaitTerminationOrTimeout(30)
```

A streaming program will start working only when you apply the start() method. And then you have to also provide a function to terminate the execution. We terminate the execution by using awaitTerminationOrTimeout() after 30 seconds. Both methods, start() and awaitTerminationOrTimeout(), are defined on StreamingContext. The output of our code is as follows:

```
-------------------------------------------
Time: 2017-08-27 12:58:20
-------------------------------------------
88.0
109.0
163.0
97.0
107.0


-------------------------------------------
Time: 2017-08-27 12:58:30
-------------------------------------------


-------------------------------------------
Time: 2017-08-27 12:58:40
-------------------------------------------
97.0
107.0
-------------------------------------------
Time: 2017-08-27 12:58:50
-------------------------------------------
107.0
162.0


-------------------------------------------
Time: 2017-08-27 12:59:00
-------------------------------------------
```

For the time interval of 10 seconds, whatever row it is getting, it returns the sum of numbers in each row.

Recipe 7-4. Integrate PySpark Streaming with Apache Kafka, and Read and Analyze the Data

Problem

You want to integrate PySpark Streaming with Apache Kafka, and read and analyze the data.

Solution

We discussed Apache Kafka in Chapter 1, and you already know the meaning of *topic*, *producer*, *consumer*, and *broker*. You know that Kafka uses ZooKeeper. Let's see the working of Apache Kafka. We will start a ZooKeeper server. Then we will create a Kafka topic. Let's start a ZooKeeper server first:

```
Starting zookeeper

kafka$ bin/zookeeper-server-start.sh config/zookeeper.properties
```

The ZooKeeper server is started by using the `zookeeper-server-start.sh` script. This script is in the `bin` directory of the Kafka installation. After starting ZooKeeper, we are going to create a Kafka topic. This can be created by using the `kafka-topics.sh` script, which resides in the `bin` directory:

```
Starting a Kafka Topic

kafka$ bin/kafka-topics.sh --create --zookeeper localhost:2185
--replication-factor 1 --partitions 1 --topic  pysparkBookTopic
```

Here is the output:

```
Created topic "pysparkBookTopic".
```

We have provided many options to Kafka. The ZooKeeper option provides data about ZooKeeper; we provide a replication factor of 1, as we are not replicating the data. On a real-time server, you are supposed to replicate the data for fault-tolerance.

We have created a topic named `pysparkBookTopic`.

Let's start the broker:

```
Starting Kafka Server

kafka$ bin/kafka-server-start.sh config/server.properties
```

We are going to start a console producer. The Apache Kafka console producer will read data from the console and produce it to the broker:

```
kafka$ bin/kafka-console-producer.sh --broker-list localhost:9092 --topic
pysparkBookTopic
```

We can start the Kafka console producer by using the `kafka-console-producer.sh` script. You will find this script under the `bin` directory of Kafka home. The console producer produces message from the console.

Let's put some data in the console where the console producer has been started:

```
20 25 25 23
21 24 21 20
20 25 25 23
21 23 21 23
```

Now it is time to start a console consumer. The console consumer will finally print the data on the console:

```
kafka$ bin/kafka-console-consumer.sh --from-beginning --zookeeper
localhost:2185 --topic pysparkBookTopic
20 25 25 23
21 24 21 20
20 25 25 23

21 23 21 23
```

Just after starting the console consumer, you will see the data produced by the console producer on the console of the console consumer.

Now that you understand the workings of Kafka, we have to rework Recipe 7-3 with some changes. This time, we'll read data from the console by using Apache Kafka, and then from Kafka we will read the data by using PySpark Streaming. After reading the data, we have to analyze it. We'll read rows of numbers and calculate the sum of the numbers of each row.

How It Works

Step 7-4-1. Starting ZooKeeper, Creating the Topic, Starting the Apache Kafka Broker and the Console Producer

```
kafka$ bin/zookeeper-server-start.sh config/zookeeper.properties
```

```
kafka$ bin/kafka-server-start.sh config/server.properties
```

```
kafka$ bin/kafka-console-producer.sh --broker-list localhost:9092 --topic
pysparkBookTopic
```

```
kafka$ bin/kafka-console-producer.sh --broker-list localhost:9092 --topic
pysparkBookTopic
32 43 45
43 54 57
32 21 32
34 54 65
```

Step 7-4-2. Starting PySpark with the spark-streaming-kafka Package

```
$ pyspark --packages org.apache.spark:spark-streaming-kafka-0-8_2.11:2.0.0
```

Step 7-4-3. Creating a Sum of Each Row of Numbers

We already created this function in the previous recipe and explained it too:

```
>>> def stringToNumberSum(data):
...        removedSpaceData = data.strip()
...        if   removedSpaceData == '' :
...            return(None)
...        splittedData =  removedSpaceData.split(' ')
...        numData = [float(x) for x in splittedData]
...        sumOfData = sum(numData)
...        return (sumOfData)

>>> dataInString = '10 10 20 '
>>> stringToNumberSum(dataInString)
40.0
```

We have tested this function too.

Step 7-4-4. Reading Data from Kafka and Getting the Sum of Each Row

The function for dealing with Kafka has been defined in the KafKaUtils class. Therefore, we first have to import it. Then we have to create the StreamingContext object:

```
>>> from pyspark.streaming.kafka import KafkaUtils
>>> from pyspark.streaming import StreamingContext
>>> bookStreamContext = StreamingContext(sc, 10)
```

Next we are going to read data from Kafka by using the createStream() method defined in KafkaUtils:

```
>>> bookKafkaStream = KafkaUtils.createStream(
                                ssc = bookStreamContext,
                                zkQuorum = 'localhost:2185',
                                groupId = 'pysparkBookGroup',
                                    topics = {'pysparkBookTopic':1}
                                    )
```

The first argument of the createStream() function is the StreamingContext object. The second argument is zkQuorum, where we provide the host machine and port of ZooKeeper. The topic is pysparkBookTopic, which we already created, and 1 in the dictionary is the replication factor of the data.

```
>>> sumedData = bookKafkaStream.map( lambda data :
stringToNumberSum(data[1]))
```

After getting the data, we run a summation on it:

```
>>> sumedData.pprint()
>>> bookStreamContext.start() ;bookStreamContext.
awaitTerminationOrTimeout(30)
```

Here is the output:

```
-------------------------------------------
Time: 2017-08-26 20:21:40
-------------------------------------------
120.0
154.0

17/08/26 20:21:44 WARN BlockManager: Block input-0-1503759104200 replicated
to only 0 peer(s) instead of 1 peers
---------------------------------- --------
Time: 2017-08-26 20:21:50
------------------------------- ------------
85.0

17/08/26 20:21:51 WARN BlockManager: Block input-0-1503759110800 replicated
to only 0 peer(s) instead of 1 peers
----------------------------------- --------
Time: 2017-08-26 20:22:00
----------------------------------- -------
153.0
```

```
--------------------------------- -----------
Time: 2017-08-26 20:22:10
--------------------------------- -------

--------------------------------- -----------
Time: 2017-08-26 20:22:20
--------------------------------- -------
```

■ **Note**   You can read more about Apache Kafka in the following blogs:

https://stackoverflow.com/questions/17205561/data-modeling-with-kafka-topics-and-partitions

https://stackoverflow.com/questions/23751708/kafka-is-zookeeper-a-must

https://stackoverflow.com/documentation/apache-kafka/1986/getting-started-with-apache-kafka#t=201709171739464528411

http://kafka.apache.org/quickstart

Recipe 7-5. Execute a PySpark Script in Local Mode

Problem

You want to execute a PySpark script in local mode.

Solution

We have written the PySpark script innerJoinInPySpark.py. The content of this code file is as follows:

```
from pyspark import SparkContext
studentData = [['si1','Robin','M'],
               ['si2','Maria','F'],
               ['si3','Julie','F'],
               ['si4','Bob',  'M'],
               ['si6','William','M']]

subjectsData = [['si1','Python'],
                ['si3','Java'],
                ['si1','Java'],
                ['si2','Python'],
```

```
                 ['si3','Ruby'],
                 ['si4','C++'],
                 ['si5','C'],
                 ['si4','Python'],
                 ['si2','Java']]
ourSparkContext = SparkContext(appName = 'innerDataJoining')
ourSparkContext.setLogLevel('ERROR')
studentRDD = ourSparkContext.parallelize(studentData, 2)
studentPairedRDD = studentRDD.map(lambda val : (val[0],[val[1],val[2]]))
subjectsPairedRDD = ourSparkContext.parallelize(subjectsData, 2)
studenSubjectsInnerJoin = studentPairedRDD.join(subjectsPairedRDD)
innerJoinedData = studenSubjectsInnerJoin.collect()
print innerJoinedData
```

This is the inner join program we used in Chapter 5 for inner joins. But two extra lines have been added. The extra lines are as follows:

```
from pyspark import SparkContext
```

and

```
ourSparkContext = SparkContext(appName = 'innerDataJoining')
```

We found that in the PySpark console, PySpark itself creates the SparkContext object as sc and enables us to use it. But in PySpark scripts, we have to create our own SparkContext. SparkContext is a way to use the APIs provided by PySpark. Therefore, there are two extra lines. The first line imports SparkContext, and the second line creates the SparkContext object with the application name (appName) innerDataJoining. Let's run this PySpark script in PySpark local mode.

How It Works

To run the PySpark script, we use the spark-submit command. In the following command, local[2] means we are using two threads for execution. The spark-submit option master defines which master we are going to use. We are using the local master.

[pysparkbook@localhost bookCode]$ spark-submit - -master local[2] innerJoinInPySpark.py

Here is the output:

```
17/08/29 12:52:09 INFO executor.Executor: Starting executor ID driver on
host localhost
17/08/29 12:52:09 INFO util.Utils: Successfully started service 'org.apache.
spark.network.netty.NettyBlockTransferService' on port 42886.
17/08/29 12:52:09 INFO netty.NettyBlockTransferService: Server created on 42886
17/08/29 12:52:09 INFO storage.BlockManagerMaster: Trying to register
BlockManager
```

```
17/08/29 12:52:09 INFO storage.BlockManagerMasterEndpoint: Registering block
manager localhost:42886 with 517.4 MB RAM, BlockManagerId(driver, localhost,
42886)
17/08/29 12:52:09 INFO storage.BlockManagerMaster: Registered BlockManager
```

```
[('si3', (['Julie', 'F'], 'Java')), ('si3', (['Julie', 'F'], 'Ruby')),
('si2', (['Maria', 'F'], 'Python')), ('si2', (['Maria', 'F'], 'Java')),
('si1', (['Robin', 'M'], 'Python')), ('si1', (['Robin', 'M'], 'Java')),
('si4', (['Bob', 'M'], 'C++')), ('si4', (['Bob', 'M'], 'Python'))]
```

We have the inner join data as output. In the next recipe, we are going to run the same script using Standalone and Mesos cluster managers.

Recipe 7-6. Execute a PySpark Script Using Standalone Cluster Manager and Mesos Cluster Manager
Problem

You want to execute a PySpark script by using Standalone Cluster Manager and the Mesos cluster manager.

Solution

We can execute our script by using spark-submit. But first we have to start Standalone Cluster Manager. It can be started with the start-all.sh script in sbin of SparkHome:

```
[pysparkbook@localhost sbin]$ /allPySpark/spark/sbin/start-all.sh
```

Here is the output:

```
starting org.apache.spark.deploy.master.Master, logging to /allPySpark/
logSpark//spark-pysparkbook-org.apache.spark.deploy.master.Master-1-
localhost.localdomain.out
localhost: starting org.apache.spark.deploy.worker.Worker, logging to /
allPySpark/logSpark//spark-pysparkbook-org.apache.spark.deploy.worker.
Worker-1-localhost.localdomain.out
```

Similarly, to run on Mesos, we have to start the Mesos master and slaves:

```
[pysparkbook@localhost bookCode]$ mesos-master --work_dir=/allPySpark/mesos/
workdir &
```

```
[root@localhost bookCode]# mesos-slave --master=127.0.0.1:5050 --work_dir=/
allPySpark/mesos/workdir --systemd_runtime_directory=/allPySpark/mesos/systemd &
```

How It Works

Let's first run our script by using Standalone Cluster Manager:

```
[pysparkbook@localhost bookCode]$ spark-submit --master spark://localhost.
localdomain:7077 --num-executors 2 innerJoinInPySpark.py
```

Here is the output:

```
[('si3', (['Julie', 'F'], 'Java')), ('si3', (['Julie', 'F'], 'Ruby')),
('si2', (['Maria', 'F'], 'Python')), ('si2', (['Maria', 'F'], 'Java')),
('si1', (['Robin', 'M'], 'Python')), ('si1', (['Robin', 'M'], 'Java')),
('si4', (['Bob', 'M'], 'C++')), ('si4', (['Bob', 'M'], 'Python'))]
```

In the command, the value of the master option is the Standalone master URL. Similarly, we can execute on Mesos by using the Mesos master URL:

```
spark-submit --master mesos://127.0.0.1:5050 --conf spark.executor.uri=/home/
pysparkbook/binaries/spark-1.6.2-bin-hadoop2.6.tgz innerJoinInPySpark.py
```

CHAPTER 8

■ ■ ■

PySparkSQL

Most data that a data scientist deals with is either structured or semistructured. Nowadays, the amount of unstructured data is increasing rapidly. The *PySparkSQL* module is a higher-level abstraction over PySpark Core for processing structured and semistructured datasets. By using PySparkSQL, we can use SQL and HiveQL code too, which makes this module popular among database programmers and Apache Hive users. The APIs provided by PySparkSQL are optimized. PySparkSQL can read data from many file types such as CSV files, JSON files, and files from other databases.

The DataFrame abstraction is similar to a table in a relational database management system. The DataFrame consists of named columns and is a collection of Row objects. Row objects are defined in PySparkSQL. Users are familiar with the schema of tabular forms, so it becomes easy to operate on DataFrames.

In PySparkSQL 1.6, a new Dataset interface was included. This interface is a hybrid of the DataFrame and RDD, so it provides the benefits of both. The Dataset interface has not been implemented in Spark with Python.

The GraphFrames library is used to process graphs. It is similar to the GraphX library, which does not work for Python. For PySpark users, the GraphFrames library is most suitable for graph processing. It has been developed on top of the SparkSQL DataFrame. We can run our DataFrame queries by using GraphFrames, which makes it unique from GraphX.

You'll find this chapter full of exciting topics, from PySparkSQL and DataFrames to and the graph analysis recipes.

This chapter covers the following recipes:

Recipe 8-1. Create a DataFrame

Recipe 8-2. Perform exploratory data analysis on DataFrames

Recipe 8-3. Perform aggregation operations on DataFrames

Recipe 8-4. Execute SQL and HiveQL queries on DataFrames

Recipe 8-5. Perform data joining on DataFrames

Recipe 8-6. Calculate breadth-first searches using GraphFrames

Recipe 8-7. Calculate page rank using GraphFrames

Recipe 8-8. Read data from Apache Hive

© Raju Kumar Mishra 2018
R. K. Mishra, *PySpark Recipes*, https://doi.org/10.1007/978-1-4842-3141-8_8

Recipe 8-1. Create a DataFrame
Problem

You want to create a DataFrame.

Solution

As you know, a *DataFrame* is collection of named columns. You might remember the filament data from Chapter 5. You want to do the following on the filament data:

- Create a DataFrame

- Know the schema of a DataFrame

- Print the content of a DataFrame

- Filter out the data for 100W bulbs

- Select data from a DataFrame for bulbs of 100W with a life greater than 650

How It Works
Step 8-1-1. Creating a Nested List of Filament Data

First we have to create a nested list of filament data. You're already familiar with this data from Chapter 5:

```
>>> filamentData = [['filamentA','100W',605],
... ['filamentB','100W',683],
... ['filamentB','100W',691],
... ['filamentB','200W',561],
... ['filamentA','200W',530],
... ['filamentA','100W',619],
... ['filamentB','100W',686],
... ['filamentB','200W',600],
... ['filamentB','100W',696],
... ['filamentA','200W',579],
... ['filamentA','200W',520],
... ['filamentA','100W',622],
... ['filamentA','100W',668],
... ['filamentB','200W',569],
... ['filamentB','200W',555],
... ['filamentA','200W',541]]
```

After creating this filamentData nested list, we are going to create an RDD of it. To create the RDD, we'll use our parallelize() function:

```
>>> filamentDataRDD = sc.parallelize(filamentData, 4)
>>> filamentDataRDD.take(4)
```

Here is the output:

```
[['filamentA', '100W', 605],
 ['filamentB', '100W', 683],
 ['filamentB', '100W', 691],
 ['filamentB', '200W', 561]]
```

The filamentDataRDD RDD has four partitions. We have created our RDD successfully. The next step is to create a schema for our DataFrame.

Step 8-1-2. Creating a Schema of a DataFrame

In our DataFrame, we have three columns. First, we are going to define the columns. We define the columns by using the StructField() function. PySparkSQL has its own data types, and all of these are defined in the submodule pyspark.sql.types. We have to import everything from pyspark.sql.types:

```
>>>from pyspark.sql.types import *
```

After importing the required submodule, we define our first column of the DataFrame:

```
>>> FilamentTypeColumn = StructField("FilamentType",StringType(),True)
```

Let's look at the arguments of StructField(). The first argument is the column name. In this example, the column name is FilamentType. The second argument is the data type of the elements in the column. In this example, the data type of the first column is StringType(). We know that some elements of a column might be null. So the last argument, which has the value True, indicates that this column might have null values or missing data.

```
>>> BulbPowerColumn = StructField("BulbPower",StringType(),True)
>>> LifeInHoursColumn = StructField("LifeInHours",StringType(),True)
```

We have created a StructField of each column. Now we have to create a schema of full DataFrames by using the StructType object as follows:

```
>>> FilamentDataFrameSchema = StructType([FilamentTypeColumn,
BulbPowerColumn, LifeInHoursColumn])
```

FilamentDataFrameSchema is the full schema of our DataFrame.

```
>>> FilamentDataFrameSchema
```

Here is the output:

```
StructType(
      List(StructField(FilamentType,StringType,true),
      StructField(BulbPower,StringType,true),
      StructField(LifeInHours,StringType,true))
      )
```

The schema of the three columns of our DataFrame can be observed via the FilamentDataFrameSchema variable, as shown in the preceding code.

We know that a DataFrame is an RDD of Row objects. Therefore, we have to transform our filamentDataRDD RDD to the RDD of Row objects. In the RDD of Row objects, every row is a Row object. In the next recipe step, we are going to transform our RDD to an RDD of Row objects.

Step 8-1-3. Creating an RDD of Row Objects

The RDD map() function is best for transforming any RDD from one structure to another. In order to transform our filamentRDD2 RDD to an RDD of Row objects. A DataFrame is nothing but an RDD of Row objects. Let's create an RDD of rows. But in order to work with Row, we have to first import it. Row is in pyspark.sql. We can import Row as shown here:

```
>>> from pyspark.sql import Row

>>> filamentRDDofROWs = filamentDataRDD.map(lambda x
:Row(str(x[0]), str(x[1]), str(x[2])))

>>> filamentRDDofROWs.take(4)
```

Here is the output:

```
[<Row(filamentA, 100W, 605)>,
 <Row(filamentB, 100W, 683)>,
 <Row(filamentB, 100W, 691)>,
 <Row(filamentB, 200W, 561)>]
```

You can see that we have created an RDD of rows, filamentRDDofROWs. We apply the take() function on our RDD and print four rows of elements out of that.

Step 8-1-4. Creating a DataFrame

We have created the schema and RDD of rows. Therefore, we can create our DataFrame. In order to create a DataFrame, we need the SQLContext object. Let's create the SQLContext object in the following line of code:

```
>>> from pyspark.sql import SQLContext
>>> sqlContext = SQLContext(sc)
```

We have created our sqlContext object. As SparkContext, in our case means in PySpark console sc is entering point to PySpark I have mentioned that sc is an object of SparkContext. In a similar way, SQLContext is the entering point to PySparkSQL.

Using the createDataFrame() function, which has been defined on SQLContext, we'll create our DataFrame, filamentDataFrameRaw. We provide two arguments to the createDataFrame() function: the first one is an RDD of Row objects filamentRDDofROWs; and the second argument is the schema for our DataFrame, FilamentDataFrameSchema.

```
>>> filamentDataFrameRaw = sqlContext.createDataFrame(filamentRDDofROWs,
FilamentDataFrameSchema)
```

We have created our DataFrame from our filament data. We have given our DataFrame the reference filamentDataFrameRaw. Let's print the records of our DataFrame on the console. Previously, we used the take() function to fetch data from the DataFrame. But now we are going to change the way we fetch data; we are going to use the show() function. The show() function prints data in a beautiful way. We can provide the number of rows as input to the show() function. In the following line, four records are being fetched:

```
>>> filamentDataFrameRaw.show(4)
```

Here is the output, showing only the top four rows:

```
+------------+---------+-----------+
|FilamentType|BulbPower|LifeInHours|
+------------+---------+-----------+
|   filamentA|     100W|        605|
|   filamentB|     100W|        683|
|   filamentB|     100W|        691|
|   filamentB|     200W|        561|
+------------+---------+-----------+
```

Step 8-1-5. Printing a Schema of a DataFrame

We have created our DataFrame. Let's check its schema. We can fetch the schema of a DataFrame by using the printSchema() function defined on the DataFrame:

```
>>> filamentDataFrameRaw.printSchema()
```

Here is the output:

```
root
 |-- FilamentType: string (nullable = true)
 |-- BulbPower: string (nullable = true)
 |-- LifeInHours: string (nullable = true)
```

The printSchema() function's output shows that the DataFrame has three columns. These columns indicate the column name, the data type of the columns, and whether a column is nullable. You are an observant reader if you have noticed that the data type of the LifeInHours column is string. It's better to represent time in hours as either an integer data type or a floating-point type. Therefore, we have to change the data type of the third column.

We can typecast a column value from one data type to another by using the cast() function.

Step 8-1-6. Changing the Data Type of a Column

The withColumn() function returns a DataFrame by adding a new column to it. But if that column is already in the DataFrame, the withColumn() function will replace the existing column:

```
>>> filamentDataFrame = filamentDataFrameRaw.withColumn('LifeInHours',filame
ntDataFrameRaw.LifeInHours.cast(FloatType()))
```

Investigating schema will now return float as the data type for the LifeInHours column:

```
>>> filamentDataFrame.printSchema()
```

Here is the output:

```
root
 |-- FilamentType: string (nullable = true)
 |-- BulbPower: string (nullable = true)
 |-- LifeInHours: float (nullable = true)
```

The data type of the LifeInHours column has been transformed from a string type to a float type. Let's investigate the DataFrame by fetching some rows from it. We can display rows by using the following function:

```
>>> filamentDataFrame.show(5)
```

Here is the output, showing only the top five rows:

```
+--------------+----------+------------+
|FilamentType  |BulbPower |LifeInHours |
+--------------+----------+------------+
|   filamentA  |   100W   |   605.0    |
|   filamentB  |   100W   |   683.0    |
|   filamentB  |   100W   |   691.0    |
|   filamentB  |   200W   |   561.0    |
|   filamentA  |   200W   |   530.0    |
+--------------+----------+------------+
```

We can observe that the data type of the LifeInHours column has been changed to floating-point.

The column names can be fetched by using the columns attribute of the DataFrame object:

```
>>> filamentDataFrame.columns
```

Here is the output:

```
['FilamentType', 'BulbPower', 'LifeInHours']
```

Step 8-1-7. Filtering Out Data Where BulbPower Is 100W

Filtering rows, based on particular logic, can be done by using the filter() function. This function takes a logical expression and returns a DataFrame of filtered data:

```
>>> filamentDataFrame100Watt = filamentDataFrame.filter(filamentDataFrame.
BulbPower == '100W')
```

We need all the rows where BulbPower is equal to 100W. Therefore, we provide filamentDataFrame.BulbPower == '100W' as an argument to the filter() function. Let's see what is inside the filamentDataFrame100Watt DataFrame:

```
>>> filamentDataFrame100Watt.show()
```

Here is the output:

```
+---------------+-----------+---------------+
|FilamentType   |BulbPower  |LifeInHours    |
+---------------+-----------+---------------+
|    filamentA  |    100W   |    605.0      |
|    filamentB  |    100W   |    683.0      |
|    filamentB  |    100W   |    691.0      |
|    filamentA  |    100W   |    619.0      |
|    filamentB  |    100W   |    686.0      |
|    filamentB  |    100W   |    696.0      |
|    filamentA  |    100W   |    622.0      |
|    filamentA  |    100W   |    668.0      |
+---------------+-----------+---------------+
```

The filter() function has done its job accurately.

Step 8-1-8. Selecting Data from a DataFrame

A compound logical expression can also be used in the filter() function. In this step, we are going to use a compound logical expression with the & operator:

```
>>> filamentData100WGreater650 =filamentDataFrame.filter((filamentDataFrame.
BPower == '100W')  & (filamentDataFrame.LifeInHours > 650.0))
```

```
>>> filamentData100WGreater650.show()
```

Here is the output:

```
+--------------+-----------+------------+
|FilamentType  |BulbPower  |LifeInHours |
+--------------+-----------+------------+
|   filamentB  |    100W   |    683.0   |
|   filamentB  |    100W   |    691.0   |
|   filamentB  |    100W   |    686.0   |
|   filamentB  |    100W   |    696.0   |
|   filamentA  |    100W   |    668.0   |
+--------------+-----------+------------+
```

Finally, we have met our requirement. In the next recipe, we are going to do exploratory analysis on a DataFrame.

Recipe 8-2. Perform Exploratory Data Analysis on a DataFrame

Problem

You want to perform exploratory data analysis on a DataFrame.

Solution

In exploratory data analysis, we explore the given data. *Exploring* the data means counting the number of records and then looking for meaningful patterns. For data of numeric columns, we calculate the measures of central tendency and the spread in the data. The *spread* in the data is nothing but the variability in the data. You might know that *measures of central tendency* are the mean, median, and mode. But how is variability, or data spread, measured? We can measure it by using either variance or standard deviation.

A given dataset might have categorical columns. For categorical columns, we count the frequency for each value of that variable. A count of records gives us an idea of the number of data points we have. We calculate the minimum and maximum data points from a given numerical column. Knowing the minimum and maximum shows us the range of data.

PySparkSQL has a summary() function defined on the DataFrame. This function will return the number of records (count), mean, standard deviation (stdev), minimum (min), and maximum (max) from a column of numerical values in the DataFrame.

You have a file filamentData.csv. This time we have to read data from the CSV file and create a DataFrame. After creating the DataFrame, we have to do a summary analysis on the DataFrame's numerical columns. Apart from summary statistics on the numerical columns, we have to know the frequency of distinct values in each categorical field.

You want to perform the following on the DataFrame of filament data:

- Read data from the CSV file filamentData.csv

- Create a DataFrame

- Calculate summary statistics on a numerical column

- Count the frequency of distinct values in the FilamentType categorical column

- Count the frequency of distinct values in the BulbPower categorical column

How It Works

First we have to read the given file and transform the data into a DataFrame. In the preceding recipe, we started from a nested list and performed several steps to create a DataFrame. And in Chapter 6 we found that it took numerous steps to get a nested list from a CSV file. It will be good if there is some.

The PySpark package can read the CSV file and transform data directly to a DataFrame. And we should be happy that we have a PySpark package to help us. The package name is com.databricks.spark.csv. This package was developed by Databricks. Before PySpark.2.x.x, the user had to use this package separately. But in PySpark version 2.x.x.x, the package is merged in PySpark, so you don't need to include the JARs separately. Thanks to Databricks for this beautiful and very useful package.

Step 8-2-1. Defining the DataFrame Schema

Here we are going to define the schema of our DataFrame. In our schema, there are three columns. The first column is FilamentType, which has the data type of string. The second column is BulbPower, which also has the data type of string. The last column is LifeInHours, which has the data type of double.

We need different data types defined in PySparkSQL. We can find all the data types in the pyspark.sql.types submodule:

```
>>> from pyspark.sql.types import *
>>> FilamentTypeColumn = StructField("FilamentType",StringType(),True)
>>> BulbPowerColumn = StructField("BulbPower",StringType(),True)
>>> LifeInHoursColumn = StructField("LifeInHours",DoubleType(),True)
```

We have created three StructFields:

```
>>> FilamentDataFrameSchema = StructType([FilamentTypeColumn,
BulbPowerColumn, LifeInHoursColumn])
```

Using these StructFields, we have created a schema for our DataFrame. The name of our DataFrame schema is FilamentDataFrameSchema.

Step 8-2-2. Reading a CSV File and Creating a DataFrame

Let's create a DataFrame. We are going use a spark.read.csv function to read and convert the file data to a DataFrame:

```
>>> filamentDataFrame = spark.read.csv('file:///home/pysparkbook/
pysparkBookData/filamentData.csv',header=True, schema =
FilamentDataFrameSchema, mode="DROPMALFORMED")
```

Let's discuss the arguments of the spark.read.csv function. The first argument is the file path. The second argument indicates that our file has a header line. The third argument provides the schema of our DataFrame. What is the mode argument? This fourth argument, mode, provides a way to deal with corrupt records during parsing. The value of the mode argument is DROPMALFORMED. This value is saying to drop all the corrupt data.

We have read our filamentData.csv data file. And the spark.read.csv() function has already transformed our CSV data into a DataFrame. Let's check whether we have our DataFrame. Do you remember the functions on a DataFrame that can help you visualize it? The show() function will work for our purpose. Let's fetch five rows from our DataFrame. The show() function prints records of the DataFrame on the console. It prints records in a tabular format, which is easier to read and understand. Let's apply the show() function to print five records on the console:

```
>>> filamentDataFrame.show(5)
```

Here is the output, showing only the top five rows:

```
+------------+---------+-----------+
|FilamentType|BulbPower|LifeInHours|
+------------+---------+-----------+
|   filamentA|     100W|      605.0|
|   filamentB|     100W|      683.0|
|   filamentB|     100W|      691.0|
|   filamentB|     200W|      561.0|
|   filamentA|     200W|      530.0|
+------------+---------+-----------+
```

We have our DataFrame. Let's check its schema:

```
>>> filamentDataFrame.printSchema()
```

Here is the output:

```
root
 |-- FilamentType: string (nullable = true)
 |-- BulbPower: string (nullable = true)
 |-- LifeInHours: double (nullable = true)
```

We have a proper schema too.

■ **Note**   To learn more about getting a CSV file into a Spark DataFrame, read this Stack Overflow discussion: https://stackoverflow.com/questions/29936156/get-csv-to-spark-dataframe.

Step 8-2-3. Calculating Summary Statistics

The describe() function, which is defined on a DataFrame, will give us the following:

```
>>> dataSummary = filamentDataFrame.describe()
>>> dataSummary.show()
```

Here is the output:

```
+---------+----------------------------+
|summary  |        LifeInHours         |
+---------+----------------------------+
|  count  |            16              |
|   mean  |          607.8125          |
| stddev  | 61.11652122517009          |
|    min  |           520.0            |
|    max  |           696.0            |
+---------+----------------------------+
```

We have our results: count, mean, stddev, min, and max. Next, we have to find the frequency of the values of our two categorical columns, FilamentType and BulbPower, and a combination of them.

Step 8-2-4. Counting the Frequency of Distinct Values in the FilamentType Categorical Column

A very naive method for finding the frequency of values is to filter the records by using the filter() function and then count them. Let's perform these tasks one by one:

```
>>> filamentDataFrame.filter(filamentDataFrame.FilamentType == 'filamentA').
count()
```

Here is the output:

8

In the preceding code, we filter out all the records where FilamentType is equal to filamentA. Eight rows have filamentA. Now let's see how many rows have filamentB in the first column:

```
>>> filamentDataFrame.filter(filamentDataFrame.FilamentType == 'filamentB').
count()
```

Here is the output:

8

We have filtered out all the records where FilamentType is equal to filamentB. Using the count() function on the filtered data returns the total number of rows that have filamentB in the first column. Eight rows have filamentB in the first column.

Step 8-2-5. Counting the Frequency of Distinct Values in the BulbPower Categorical Column

Now let's filter data where BulbPower is equal to 100W. After filtering out the required rows, we have to count them:

```
>>> filamentDataFrame.filter(filamentDataFrame.BulbPower == '100W').count()
```

Here is the output:

```
8
```

Eight rows have BulbPower values of 100W. Similarly, we can count other values and their combinations. Let's compute the frequency of 200W bulbs:

```
>>> filamentDataFrame.filter(filamentDataFrame.BulbPower == '200W').count()
```

Here is the output:

```
8
```

Our BulbPower columns have eight 100W bulbs and eight 200W bulbs.

Step 8-2-6. Counting the Frequency of Distinct Values in a Combination of FilamentType and BulbPower Columns

In the following code, we are going to count rows on the basis of a compound logical expression:

```
>>> filamentDataFrame.filter((filamentDataFrame.FilamentType == 'filamentB')
& (filamentDataFrame.BulbPower == '100W')).count()
```

Here is the output:

```
4
>>> filamentDataFrame.filter((filamentDataFrame.FilamentType == 'filamentB')
& (filamentDataFrame.BulbPower == '200W')).count()
```

Here is the output:

```
4
```

```
>>> filamentDataFrame.filter((filamentDataFrame.FilamentType == 'filamentA')
& (filamentDataFrame.BulbPower == '200W')).count()
```

Here is the output:

4

```
>>> filamentDataFrame.filter((filamentDataFrame.FilamentType == 'filamentA')
& (filamentDataFrame.BulbPower == '100W')).count()
```

Here is the output:

4

Recipe 8-3. Perform Aggregation Operations on a DataFrame

Problem

You want to perform data aggregation on a DataFrame.

Solution

To get a summarized pattern of data, data scientists perform aggregation on a given dataset. Summarized patterns are easy to understand. Sometimes the summarization is done based on the key. To perform aggregation based on the key, we first need to group the data by key.

In PySparkSQL, grouping by key can be performed by using the groupBy() function. This function returns the pyspark.sql.group.GroupedData object. After this GroupedData object is created, we can apply many aggregation functions such as avg(), sum(), count(), min(), max(), and sum() on GroupedData.

We have a data file named adult.data. I obtained this data from the web site of the Bren School of Information and Computer Science at the University of California, Irvine (UCI). This is a simple CSV file with 15 columns. Table 8-1 describes all 15 columns.

You want to do the following:

- Create a DataFrame from the adult.data file

- Count the total number of records in the DataFrame

- Count the number of times that a salary is greater than $50,000 and the number of times it's less than $50,000

- Perform summary statistics on the numeric columns age, capital-gain, capital-loss, and hours-per-week

- Find out the mean age of male and female workers from the data

- Find out whether a salary greater than $50,000 is more frequent for males or females

- Find the highest-paid job

Table 8-1. *Description of adult.data File*

Columns	Description
age	Age of person, continuous
workclass	Private, Self-emp-not-inc, Self-emp-inc, Federal-gov, Local-gov, State-gov, Without-pay, Never-worked.
fnlwgt	Continuous
education	Bachelors, Some-college, 11th, HS-grad, Prof-school, Assoc-acdm, Assoc-voc, 9th, 7th-8th, 12th, Masters, 1st-4th, 10th, Doctorate, 5th-6th, Preschool.
Education-num	Continuous
Marital-status	Married-civ-spouse, Divorced, Never-married, Separated, Widowed, Married-spouse-absent, Married-AF-spouse.
occupation	Tech-support, Craft-repair, Other-service, Sales, Exec-managerial, Prof-sepcialty, Handlers-cleaners, Machine-op-inspct, Adm-clerical, Farming-fishing, Transport-moving, Priv-house-serv, Protective-serv, Armed-Forces.
relationship	Relationship: Wife, Own-child, Husband, Not-in-family, Other-relative, Unmarried.
race	White, Asian-Pac-Islander, Amer-Indian-Eskimo, Other, Black.
sex	Female, Male.
Capital-gain	Continuous
Capital-loss	Continuous
Hours-per-week	Continuous
Native-country	United-States, Cambodia, England, Puerto-Rico, Canada, Germany, Outlying-US(Guam-USVI-etc), India, Japan, Greece, South, China, Cuba, Iran, Honduras, Philippines, Italy, Poland, Jamaica, Vietnam, Mexico, Portugal, Ireland, France, Dominican-Republic, Laos, Ecuador, Taiwan, Haiti, Columbia, Hungary, Guatemala, Nicaragua, Scotland, Thailand, Yugoslavia, El-Salvador, Trinadad&Tobago, Peru, Hong, Holand-Netherlands.
Class (income)	>50K, <=50K

How It Works

First we will download the required data, and then we will perform all the required actions one by one.

Step 8-3-1. Creating a DataFrame from the adult.data File

Let's download the file adult.data from the UCI Machine Learning Repository web site. We can fetch the data file by using the wget Linux command:

$wget https://archive.ics.uci.edu/ml/machine-learning-databases/adult/adult.data

We have download the data file. Now we will read the data by using the spark.read.csv() function:

```
>>> censusDataFrame = spark.read.csv('file:///home/pysparkbook/
pysparkBookData/adult.data',header=True, inferSchema = True)
```

In the preceding code, the second argument of the spark.read.csv() function is header = True. This indicates that the adult.data file has a header. The inferSchema = True argument is to infer the schema from the data itself. We are not providing an explicit schema for our DataFrame.

```
>>> censusDataFrame.printSchema()
```

Here is the output:

```
root
 |-- age: integer (nullable = true)
 |-- workclass: string (nullable = true)
 |-- fnlwgt: double (nullable = true)
 |-- education: string (nullable = true)
 |-- education-num: double (nullable = true)
 |-- marital-status: string (nullable = true)
 |-- occupation: string (nullable = true)
 |-- relationship: string (nullable = true)
 |-- race: string (nullable = true)
 |-- sex: string (nullable = true)
 |-- capital-gain: double (nullable = true)
 |-- capital-loss: double (nullable = true)
 |-- hours-per-week: double (nullable = true)
 |-- native-country: string (nullable = true)
 |-- income: string (nullable = true)
```

From the schema of our DataFrame, it is clear that there are 15 columns. Some columns are numeric, and the rest are strings. Our dataset is a mixture of categorical and numerical fields.

Step 8-3-2. Counting the Total Number of Records in a DataFrame

Let's count how many records we have in our DataFrame. Our simplest count() method fulfills that requirement:

```
>>> censusDataFrame.count()
```

Here is the output:

```
32561
```

Our data frame has 32,561 records. That's a large number of records.

Step 8-3-3. Counting the Frequency of Salaries Greater Than and Less Than 50K

We can achieve our goal of counting the frequency of certain salaries by first grouping our data by the income column and then counting by using our count() function:

```
>>> groupedByIncome = censusDataFrame.groupBy('income').count()
>>> groupedByIncome.show()
```

Here is the output:

```
+----------+----------+
|income    |count     |
+----------+----------+
|   >50K   |  7841    |
|  <=50K   | 24720    |
+----------+----------+
```

It is evident from this table that salaries greater than $50,000 are less frequent than those less than or equal to that amount.

Step 8-3-4. Performing Summary Statistics on Numeric Columns

The describe() function, shown here, is very useful:

```
>>> censusDataFrame.describe('age').show()
```

Here is the output:

```
+-------+------------------+
|summary|               age|
+-------+------------------+
|  count|             32561|
|   mean| 38.58164675532078|
| stddev|13.640432553581356|
|    min|                17|
|    max|                90|
+-------+------------------+
```

The maximum age of a working person is 90 years old, and the minimum age is 17 years. The mean age of working people is 30.58 years. From this mean value, it can be inferred that most working people are in their 30s.

Similarly, we can find summary statistics for the capital-gain and capital-loss data:

```
>>> censusDataFrame.describe('capital-gain').show()
```

```
+-------+------------------+
|summary|      capital-gain|
+-------+------------------+
|  count|             32561|
|   mean|1077.6488437087312|
| stddev| 7385.292084840354|
|    min|               0.0|
|    max|           99999.0|
+-------+------------------+
```

```
>>> censusDataFrame.describe('capital-loss').show()
```

Here is the output:

```
+-------+----------------+
|summary|    capital-loss|
+-------+----------------+
|  count|           32561|
|   mean| 87.303829734959|
| stddev|402.960218649002|
|    min|             0.0|
|    max|          4356.0|
+-------+----------------+
```

Let's see the distribution of hours per week for workers:

```
>>> censusDataFrame.describe('hours-per-week').show()
```

Here is the output:

```
+-------+------------------+
|summary|    hours-per-week|
+-------+------------------+
|  count|             32561|
|   mean|40.437455852092995|
| stddev|12.347428681731838|
|    min|               1.0|
|    max|              99.0|
+-------+------------------+
```

It is clear that the maximum number of working hours per week is 99.0.

Step 8-3-5. Finding the Mean Age of Male and Female Workers from the Data

The average age of male and female workers can be found as follows:

```
>>> groupedByGender = censusDataFrame.groupBy('sex')

>>> type(groupedByGender)
```

Here is the output:

```
<class 'pyspark.sql.group.GroupedData'>

>>> groupedByGender.mean('age').show()
```

Here is the output:

```
+----------+----------------------+
|   sex    |      avg(age)        |
+----------+----------------------+
|   Male   |  39.43354749885268   |
|  Female  |  36.85823043357163   |
+----------+----------------------+
```

Male workers are, on average, older than their female worker counterparts.

```
>>> groupedByGender.mean('hours-per-week').show()
```

Here is the output:

```
+----------+----------------------------+
|   sex    |    avg(hours-per-week)     |
+----------+----------------------------+
|   Male   |    42.42808627810923       |
|  Female  |    36.410361154953115      |
+----------+----------------------------+
```

Female workers, on average, work fewer hours than their male counterparts, according to our data.

Step 8-3-6. Finding Out Whether High Salaries are More Frequent for Males or Females

Since our result depends on two fields, sex and income, we have to group our data by both:

```
>>> groupedByGenderIncome = censusDataFrame.groupBy(['income', 'sex'])
```

Now, on the grouped data, we can apply the count() function to get our desired result:

```
>>> groupedByGenderIncome.count().show()
```

Here is the output:

```
+------+-------+-----+
|income|    sex|count|
+------+-------+-----+
|  >50K|   Male| 6662|
|  >50K| Female| 1179|
| <=50K| Female| 9592|
| <=50K|   Male|15128|
+------+-------+-----+
```

We can see that a salary greater than $50,000 is more frequent for males.

Step 8-3-7. Finding the Highest-Paid Job

To find the job with highest income, we must group our DataFrame on the occupation and income fields:

```
>>> groupedByOccupationIncome = censusDataFrame.groupBy(['occupation',
'income'])
```

On the grouped data, we have to apply count(). We need the highest-paid occupation, so we need to sort the data:

```
>>> groupedByOccupationIncome.count().sort(['income','count'],
ascending= 0).show(5)
```

Here is the output, showing only the top five rows:

```
+---------------+------+-----+
|     occupation|income|count|
+---------------+------+-----+
| Exec-managerial|  >50K| 1968|
|  Prof-specialty|  >50K| 1859|
|           Sales|  >50K|  983|
|     Craft-repair|  >50K|  929|
|     Adm-clerical|  >50K|  507|
+---------------+------+-----+
```

We can see that a high frequency of executive/managerial people have salaries greater than $50,000.

Recipe 8-4. Execute SQL and HiveQL Queries on a DataFrame

Problem

You want to run SQL and HiveQL queries on a DataFrame.

Solution

We can use `createOrReplaceTempView()`, which creates a temporary view. The `DataFrame` class provides this function. The life of this view is the same as the `SparkSession` that creates the DataFrame.

We have another function, `registerTempTable()`, which creates a temporary table in memory. Using `SQLContext`, we can run SQL commands, and using `HiveContext`, we can run HiveQL queries on these temporary tables. In new versions of PySpark, this method is deprecated. But if you are working with older PySpark code, you might find `registerTempTable()`.

In the preceding recipe, we created a DataFrame named `censusDataFrame`. You want to perform the following actions on the DataFrame:

- Create a temporary view

- Select the age and `income` columns by using SQL commands

- Compute the average hours worked per week, based on education level

How It Works

We will start with creating a temporary view of our DataFrame. After creating this temporary view, we will apply SQL commands to perform our tasks.

Step 8-4-1. Creating a Temporary View in Memory

Let's create a temporary table first. Then we can run our SQL or HiveQL commands on that table.

```
>>> censusDataFrame.createOrReplaceTempView("censusDataTable")
```

This creates the temporary table `censusDataTable`.

Step 8-4-2. Selecting Age and Income Columns Using a SQL Command

The select command is a highly used and very popular command in SQL. We need to select two columns, age and income:

```
>>> censusDataAgeIncome = spark.sql('select age, income from censusDataTable
limit 5')
```

The SQL command in the preceding code is a general SQL select command to fetch two columns, age and income. The spark.sql() function can be used to run SQL commands.

In the spark.sql() function, Spark is denoting a SparkSession object. Whenever we start the PySpark shell, we find the SparkSession available as spark. In the preceding SQL command, limit 5 will return five records only.

Running the SQL command by using the spark.sql() function will return a DataFrame. In our case, we have the DataFrame censusDataAgeIncome, which has only two columns, age and income. So we know that we can print our DataFrame columns by using the show() function. Let's see the result:

```
>>> censusDataAgeIncome.show()
```

Here is the output:

```
+---+------+
|age|income|
+---+------+
| 39| <=50K|
| 50| <=50K|
| 38| <=50K|
| 53| <=50K|
| 28| <=50K|
+---+------+
```

The spark.sql() function returns a DataFrame. We can test this by using the type() function. The following code line is for testing the data type of censusDataAgeIncome:

```
>>> type(censusDataAgeIncome)
```

Here is the output:

```
<class 'pyspark.sql.dataframe.DataFrame'>
```

The type of censusDataAgeIncome is DataFrame.

Step 8-4-3. Computing Average Hours per Week Based on Education Level

Computing the average hours per week for workers, based on their education level, requires grouping data by education first. In SQL and HiveQL, we can use the group by clause to get the grouped data:

```
>>> avgHoursPerWeekByEducation = spark.sql("select education,
round(avg(`hours-per-week`),2) as averageHoursPerWeek from censusDataTable
group by education")
```

The SQL avg() function will find the mean, or average, value. The round() function has been used to get data in a beautifully formatted fashion. We fetch the average hours per week as an aliased name, averageHoursPerWeek. And at the end of the computation, we get the DataFrame avgHoursPerWeekByEducation, which consists of two columns. The first column is education, and the second column is averageHoursPerWeek. Let's apply the show() function to see the content of the DataFrame:

```
>>> avgHoursPerWeekByEducation.show()
```

Here is the output:

education	averageHoursPerWeek
Prof-school	47.43
10th	37.05
7th-8th	39.37
5th-6th	38.90
Assoc-acdm	40.50
Assoc-voc	41.61
Masters	43.84
12th	35.78
Preschool	36.65
9th	38.04
Bachelors	42.61
Doctorate	46.97
HS-grad	40.58
11th	33.93
Some-college	38.85
1st-4th	38.26

Recipe 8-5. Perform Data Joining on DataFrames
Problem

You want to perform join operations on two DataFrames.

Solution

Often we're required to combine information from two or more DataFrames or tables. To do this, we perform a join of DataFrames. Basically, *table joining* is a SQL term, where we join two or more tables to get denormalized tables. Join operations on two tables are very common in data science.

In PySparkSQL, we can perform the following types of joins:

- Inner join
- Left outer join
- Right outer join
- Full outer join

In Chapter 5, we discussed paired RDD joins. In this recipe, we'll discuss joining DataFrames. You want to perform the following:

- Read a student data table from a PostgreSQL database
- Read subject data from a JSON file
- Perform an inner join on DataFrames
- Save an inner-joined DataFrame as a JSON file
- Perform a right outer join
- Save a right-outer-joined DataFrame into PostgreSQL
- Perform a left outer join
- Perform a full outer join

In order to join the two DataFrames, PySparkSQL provides the `join()` function, which works on DataFrames.

How It Works

Let's start exploring the PostgreSQL database and the tables inside. We have to read data from the PostgreSQL database. Let's create a table in the PostgreSQL database and put records into the table.

Let's enter into the database server:

```
$ sudo -u postgres psql
```

For our recipe, we'll create the database pysparkbookdb:

```
postgres=# create database pysparkbookdb;
```

Here is the output:

```
CREATE DATABASE
```

We can see whether our database has been created successfully. The following SQL select command fetches all the databases that exist on our database server:

```
postgres=# SELECT datname FROM pg_database;
```

Outcome :

```
 datname
---------------
 template1
 template0
 postgres
 metastore
 pymetastore
 pysparkbookdb
(6 rows)
```

We can see that our pysparkbookdb database has been created successfully.

After creating the pysparkbookdb database, we have to connect to that database. We can connect to a database by using the \c command, which stands for *connect*. After connecting to the database, we are going to create a table, studentTable. Then we will insert our student data into the table.

```
postgres=# \c pysparkbookdb
```

You are now connected to the pysparkbookdb database as the user postgres. Let's create our required table too:

```
pysparkbookdb=# create table studentTable(
pysparkbookdb(# studentID char(50) not null,
pysparkbookdb(# name char(50) not  null,
pysparkbookdb(# gender char(5) not null
pysparkbookdb(# );
```

```
CREATE TABLE
```

We have created a studentTable table in the RDBMS. The \d, if used with the table name, provides the schema of the table; but if the command is used without anything, it prints all the tables in that particular database. In the following lines. we are printing the schema of the studentTable table:

```
pysparkbookdb=# \d  studentTable
```

Here is the output:

```
Table "public.studenttable"

 Column    |     Type      | Modifiers
-----------+---------------+-----------
 studentid | character(50) | not null
 name      | character(50) | not null
 gender    | character(5)  | not null
```

Let's put some data into the table. We are going to put in five records of students:

```
insert into studentTable values ('si1', 'Robin', 'M');
insert into studentTable values ('si2', 'Maria', 'F');
insert into studentTable values ('si3', 'Julie',  'F');
insert into studentTable values ('si4', 'Bob',    'M');
insert into studentTable values ('si6','William','M');
```

Records have been inserted into the table. We can visualize table data by using the SQL select command.

```
pysparkbookdb=# select * from studentTable;
```

Here is the output:

```
       studentid          |               name               | gender
--------------------------+----------------------------------+--------
           si1            |              Robin               |   M
           si2            |              Maria               |   F
           si3            |              Julie               |   F
           si4            |              Bob                 |   M
           si6            |              William             |   M
```

(5 rows)

We have created a table and inserted records.

Step 8-5-1. Reading Student Data Table from PostgreSQL Database

We know that we have our student data in a table in a PostgreSQL server. We need to read data from studentTable, which is in the pysparkbookdb database. In order to connect PySpark to the PostgreSQL server, we need a database JDBC connector. We are going to start our PySpark shell by using the following command:

```
pyspark --driver-class-path  .ivy2/jars/org.postgresql_postgresql-9.4.1212.
jar --packages org.postgresql:postgresql:9.4.1212
```

After starting the PySpark shell, including our connector JARs, we can read the table data by using the spark.read function:

```
>>> dbURL="jdbc:postgresql://localhost/pysparkbookdb?user=postgres&passwo
rd=''"
```

We have created our database URL. We connect to our pysparkbookdb database by using the postgres user. There is no password for the postgres user. The PostgreSQL server is running on the localhost machine. We are going to read data from the PostgreSQL server by using the spark.read function.

```
>>> studentsDataFrame = spark.read.format('jdbc').options(
url = dbURL,
database='pysparkbookdb',
dbtable='studenttable'
                                         ).load()
```

In the options part, we provide the URL of the database, the database name for the database argument, and the table name for the dbtable argument. We read the table data, which has been transformed into the DataFrame studentsDataFrame. Let's check our studentsDataFrame:

```
>>> studentsDataFrame.show()
```

Here is the output:

```
+-------------------+--------------------+------+
|          studentid|           name|gender|
+-------------------+--------------------+------+
|si1            ...|Robin         ...|  M   |
|si2            ...|Maria         ...|  F   |
|si3            ...|Julie         ...|  F   |
|si4            ...|Bob           ...|  M   |
|si6            ...|William       ...|  M   |
+-------------------+--------------------+------+
```

We have our required student DataFrame. But do you see any problems with our DataFrame? Have another look. Do you see the ellipses (...) in our DataFrame? We have to remove them. We can do this by using the trim() function, applying it to the columns.

To trim strings, we have to import the trim() function. This function is in the submodule pyspark.sql.functions. After importing the trim() function, we can use it to remove the dots from our columns as follows:

```
>>> from pyspark.sql.functions import trim
>>> studentsDataFrame = studentsDataFrame.select(trim(studentsData
Frame.studentid),trim(studentsDataFrame.name),studentsDataFrame.gender)
```

Let's print and see whether we got rid of the problem:

```
>>> studentsDataFrame.show()
```

Here is the output:

```
+---------------+---------+------+
|trim(studentid)|trim(name)|gender|
+---------------+---------+------+
|            si1|    Robin| M    |
|            si2|    Maria| F    |
|            si3|    Julie| F    |
|            si4|      Bob| M    |
|            si6|  William| M    |
+---------------+---------+------+
```

We got rid of the problem. But are you sure that we got rid of all the problems? How about the names of the DataFrame columns? Now we have to change the column names to be clearer:

```
>>> studentsDataFrame = studentsDataFrame.withColumnRenamed
('trim(studentid)', 'studentID').withColumnRenamed('trim(name)','Name').
withColumnRenamed('gender', 'Gender')
```

You can see the changed column names by printing the schema:

```
>>> studentsDataFrame.printSchema()
```

Here is the output:

```
root
 |-- studentID: string (nullable = false)
 |-- Name: string (nullable = false)
 |-- Gender: string (nullable = false)
```

We have our column names in a readable format. We should check that everything is appropriate by printing the DataFrame:

```
>>> studentsDataFrame.show()
```

Here is the output:

```
+---------+-------+------+
|studentID|   Name|Gender|
+---------+-------+------+
|      si1|  Robin| M    |
|      si2|  Maria| F    |
|      si3|  Julie| F    |
|      si4|    Bob| M    |
|      si6|William| M    |
+---------+-------+------+
```

Now we can move on to DataFrame joining.

Step 8-5-2. Reading Subject Data from a JSON File

Let's read our subject data from the subjects.json file:

```
>>> subjectsDataFrame = sqlContext.read.format("json").load('/home/
pysparkbook/pysparkBookData/subjects.json')
```

We have another DataFrame, subjectsDataFrame. Let's investigate our subjectsDataFrame by using the show() function:

```
>>> subjectsDataFrame.show()
```

Here is the output:

```
+---------+-------+
|studentID|subject|
+---------+-------+
|      si1| Python|
|      si3|   Java|
|      si1|   Java|
|      si2| Python|
|      si3|   Ruby|
|      si4|    C++|
|      si5|      C|
|      si4| Python|
|      si2|   Java|
+---------+-------+
```

```
>>> subjectsDataFrame.printSchema()
```

Here is the output:

```
root
 |-- studentID: string (nullable = true)
 |-- subject: string (nullable = true)
```

Step 8-5-3. Performing an Inner Join on DataFrames

We have two DataFrames, subjectsDataFrame and studentsDataFrame. In both DataFrames, we have to perform a join on the studentID column. This column is common in both DataFrames. An inner join returns records that have key values that match. If we look at the values of the studentID column in both DataFrames, we will find that values si1, si2, si3, and si4 are common to both DataFrames. Therefore, an inner join will return records for only those values.

```
>>> joinedDataInner = subjectsDataFrame.join(studentsDataFrame,
subjectsDataFrame.studentID==studentsDataFrame.studentID, how='inner')
```

```
>>> joinedDataInner.show()
```

Here is the output:

```
+---------+-------+---------+-----+------+
|studentID|subject|studentID| Name|Gender|
+---------+-------+---------+-----+------+
|      si1|   Java|      si1|Robin|     M|
|      si1| Python|      si1|Robin|     M|
|      si2|   Java|      si2|Maria|     F|
|      si2| Python|      si2|Maria|     F|
|      si3|   Ruby|      si3|Julie|     F|
|      si3|   Java|      si3|Julie|     F|
|      si4| Python|      si4|  Bob|     M|
|      si4|    C++|      si4|  Bob|     M|
+---------+-------+---------+-----+------+
```

In the resulting DataFrame joinedDataInner, it is easily observed that we have student IDs si1, si2, si3, and si4.

Step 8-5-4. Saving an Inner-Joined DataFrame as a JSON File

After doing analysis, we generally save the results somewhere. Here we are going to save our DataFrame joinedDataInner as a JSON file. Let's have a look at the columns in joinedDataInner; we can see that the studentID column occurs twice. If we are saving data in the same format, it is going to throw a pyspark.sql.utils.AnalysisException exception. Therefore, we first have to remove the duplicate column. For this, the select() function is going to be the most useful. The following code removes the duplicate studentID column:

```
>>> joinedDataInner = joinedDataInner.select(subjectsDataFrame.
studentID,'subject', 'Name', 'Gender')
```

The columns of the DataFrame are as follows:

```
>>> joinedDataInner.columns
['studentID', 'subject', 'Name', 'Gender']
```

The duplicate studentID column has been removed. The following code saves the DataFrame as a JSON file inside the innerJoinedTable directory;

```
>>> joinedDataInner.write.format('json').save('/home/muser/
innerJoinedTable')
```

We should see what has been saved under our innerJoinedTable directory:

```
innerJoinedTable$ ls
```

Here is the output:

```
part-r-00000-77838a67-4a1f-441a-bb42-4cd03be525a9.json   _SUCCESS
```

216

The ls command shows two files inside the directory. The JSON file contains our data, and the second file tells us that we have written the data successfully. Now you want to know what is inside the JSON file. The command cat is best for this job:

```
innerJoinedTable$ cat part-r-00000-77838a67-4a1f-441a-bb42-4cd03be525a9.json
```

Here is the output:

```
{"studentID":"si1","subject":"Java","Name":"Robin","Gender":"M    "}
{"studentID":"si1","subject":"Python","Name":"Robin","Gender":"M    "}
{"studentID":"si2","subject":"Java","Name":"Maria","Gender":"F    "}
{"studentID":"si2","subject":"Python","Name":"Maria","Gender":"F    "}
{"studentID":"si3","subject":"Ruby","Name":"Julie","Gender":"F    "}
{"studentID":"si3","subject":"Java","Name":"Julie","Gender":"F    "}
{"studentID":"si4","subject":"Python","Name":"Bob","Gender":"M    "}
{"studentID":"si4","subject":"C++","Name":"Bob","Gender":"M    "}
```

We have done one more job successfully.

Step 8-5-5. Performing a Left Outer Join

Here, we are going to perform a left outer join. In a left outer join, every value from the studentID column of the subjectsDataFrame DataFrame will be considered, even if it has a matching counterpart in the studentID column of the studentsDataFrame DataFrame. For the left outer join, we have to provide left_outer as the value of the how argument of the join() function.

```
>>> joinedDataLeftOuter = subjectsDataFrame.join(studentsDataFrame,
subjectsDataFrame.studentID==studentsDataFrame.studentID, how='left_outer')
>>> joinedDataLeftOuter.show()
```

Here is the output:

```
+---------+-------+---------+-----+------+
|studentID|subject|studentID| Name|Gender|
+---------+-------+---------+-----+------+
|      si5|      C|     null| null|  null|
|      si2| Python|      si2|Maria|     F|
|      si2|   Java|      si2|Maria|     F|
|      si4|    C++|      si4|  Bob|     M|
|      si4| Python|      si4|  Bob|     M|
|      si3|   Java|      si3|Julie|     F|
|      si3|   Ruby|      si3|Julie|     F|
|      si1| Python|      si1|Robin|     M|
|      si1|   Java|      si1|Robin|     M|
+---------+-------+---------+-----+------+
```

The left-outer-joined table shows that si5, which is part of the studentID column of subjectsDataFrame, is part of our joined table.

Step 8-5-6. Saving a Left-Outer-Joined DataFrame into PostgreSQL

Saving result data to a PostgreSQL database helps data analysts put results into safe hands. Other users can use the result data for many purposes. Let's save our results in a PostgreSQL database.

Again, we have to remove the duplicate column before saving the data to a PostgreSQL database:

```
>>> joinedDataLeftOuter = joinedDataLeftOuter.select(subjectsDataFrame.
studentID,'subject', 'Name', 'Gender')
```

It is a good idea to check that the data has been saved properly in the database. The \d command will print all the existing tables in the PostgreSQL database. We have already created the database pysparkbookdb on our server. In the same database, we are going to save our DataFrame:

```
pysparkbookdb=# \d
```

Here is the output:

```
            List of relations
 Schema |      Naame     | Type  |  Owner
--------+----------------+-------+----------
 public | studenttable   | table | postgres
(1 row)
```

The \d command shows that in the pysparkbookdb database, we have only one table, studenttable. Now let's save the DataFrame:

```
>>> props = { 'user' : 'postgres', 'password' : '' }
>>> joinedDataLeftOuter.write.jdbc(
...                                 url   = dbURL,
...                                 table = 'joineddataleftoutertable',
...                                 mode  = 'overwrite',
...                                 properties = props
...                                 )
```

The preceding code saves the DataFrame to the joineddataleftoutertable table. It also defines the variable dbURL. The mode argument has the value overwrite, which means it will overwrite the values if something before.

After saving data into PostgreSQL, we should check it once. Again, we are going to use the \d command to see all the tables in our pysparkbookdb database:

pysparkbookdb=# \d

Here is the output:

```
                    List of relations
 Schema |              Name              | Type  |  Owner
--------+--------------------------------+-------+----------
 public | joineddataleftoutertable       | table | postgres
 public | studenttable                   | table | postgres
(2 rows)
```

And we have our DataFrame saved in PostgreSQL as the joineddataleftoutertable table in the pysparkbookdb database. Let's check the values in the table by using the select command:

pysparkbookdb=# select * from joineddataleftoutertable;

Here is the output:

```
 studentID | subject | Name   | Gender
-----------+---------+--------+--------
 si5       | C       |        |
 si2       | Python  | Maria  | F
 si2       | Java    | Maria  | F
 si4       | C++     | Bob    | M
 si4       | Python  | Bob    | M
 si3       | Java    | Julie  | F
 si3       | Ruby    | Julie  | F
 si1       | Python  | Robin  | M
 si1       | Java    | Robin  | M
(9 rows)
```

For further use of the result data, we have already saved it in the PostgreSQL database. After the left outer join, we are going to perform a right outer join on our DataFrames.

Step 8-5-7. Performing a Right Outer Join

In a right outer join, every value of the studentID column of studentsDataFrame.

```
>>> joinedDataRightOuter = subjectsDataFrame.join(studentsDataFrame,
subjectsDataFrame.studentID==studentsDataFrame.studentID, how='right_outer')
>>> joinedDataRightOuter.show()
```

```
+---------+-------+---------+-------+------+
|studentID|subject|studentID|   Name|Gender|
+---------+-------+---------+-------+------+
|      si1|   Java|      si1|  Robin| M    |
|      si1| Python|      si1|  Robin| M    |
|      si2|   Java|      si2|  Maria| F    |
|      si2| Python|      si2|  Maria| F    |
|      si3|   Ruby|      si3|  Julie| F    |
|      si3|   Java|      si3|  Julie| F    |
|      si4| Python|      si4|    Bob| M    |
|      si4|    C++|      si4|    Bob| M    |
|     null|   null|      si6|William| M    |
+---------+-------+---------+-------+------+
```

Step 8-5-8. Performing a Full Outer Join

An outer join combines all values from the key columns:

```
>>> joinedDataOuter = subjectsDataFrame.join(studentsDataFrame,
subjectsDataFrame.studentID==studentsDataFrame.studentID, how='outer')
```

```
>>> joinedDataOuter.show()
```

Here is the output:

```
+---------+-------+---------+-------+------+
|studentID|subject|studentID|   Name|Gender|
+---------+-------+---------+-------+------+
|      si5|      C|     null|   null|  null|
|      si2| Python|      si2|  Maria| F    |
|      si2|   Java|      si2|  Maria| F    |
|      si4|    C++|      si4|    Bob| M    |
|      si4| Python|      si4|    Bob| M    |
|      si3|   Java|      si3|  Julie| F    |
|      si3|   Ruby|      si3|  Julie| F    |
|     null|   null|      si6|William| M    |
|      si1| Python|      si1|  Robin| M    |
|      si1|   Java|      si1|  Robin| M    |
+---------+-------+---------+-------+------+
```

Recipe 8-6. Perform Breadth-First Search Using GraphFrames

Problem

You want to perform a breadth-first search using GraphFrames.

Solution

A breadth-first search is a very popular algorithm that can be used to find the shortest distance between two given nodes. Figure 8-1 shows a graph that we have been given. It has seven nodes: A, B, C, D, E, F, and G.

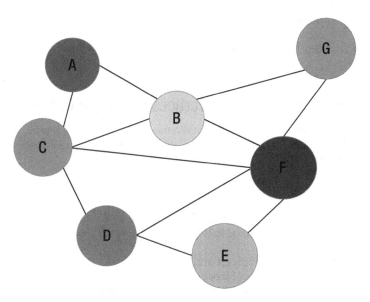

Figure 8-1. *A graph*

If we will look at the connection between nodes, we will find the following:

```
A - C - B
B - A - C - G - F
C - A - B - F - D
D - C - F - E
E - D - F
F - B - C - D -- E - G
G - B - F
```

Let me explain this structure. Take a look at the first line. This line, A - C - B, tells us that node A is connected to B and C.

You want to perform a breadth-first search and find the shortest distance between nodes. To do this, we are going to use an external library, GraphFrames. We can use PySpark with GraphFrames very easily. GraphFrames provides DataFrame-based graphs.

How It Works

To use GraphFrames, we first have to include it as we did for the PostgreSQL connector JAR file. We are going to start the PySpark shell by using the GraphFrames JAR. We are using GraphFrames version 0.4.0.

```
$ pyspark --packages graphframes:graphframes:0.4.0-spark2.0-s_2.11
```

The GraphFrames package has been added. Now we are going to run a breadth-first search using GraphFrames.

Step 8-6-1. Creating DataFrames of Vertices of a Given Graph

We know that GraphFrames work on PySparkSQL DataFrames. Let's create a DataFrame of vertices. We already know how to create a DataFrame. We are going to perform the same steps as we did in Recipe 8-1:

```
>>> from pyspark.sql.types import *
>>> from pyspark.sql import Row
>>> verticesDataList = ['A', 'B', 'C', 'D', 'E', 'F', 'G']
>>> verticesSchema = StructType([StructField('id',StringType(),True)])

>>> verticesRDD = sc.parallelize(verticesDataList, 4)
>>> verticesRDDRows = verticesRDD.map(lambda data : Row(data[0]))
>>> verticesDataFrame = sqlContext.createDataFrame(verticesRDDRows,
verticesSchema)
>>> verticesDataFrame.show(4)
```

Here is the output, showing only the top four rows:

```
+---+
| id|
+---+
|  A|
|  B|
|  C|
|  D|
+---+
```

We have created our vertices DataFrame. I am sure that you have observed that the column name of our vertices DataFrame is id. Can you have another name as the column name for the vertices DataFrame ? The answer is a simple no. It is mandatory to name the column id. Let's create a DataFrame of edges.

Step 8-6-2. Creating DataFrames of Edges of a Given Graph

We have to create a DataFrame of edges. We will first create a list of tuples; each tuple will have a source node and destination node of an edge:

```
>>> edgeDataList = [('A','C'),('A','B'),('B','A'),('B','C'),('B','G'),
                    ('B','F'),('C','A'),
                    ('C','B'),('C','F'),('C','D'),('D','C'),('D','F'),
                    ('D','E'),('E','D'),
                    ('E','F'),('F','B'),('F','C'),('F','D'),('F','E'),
                    ('F','G'),('G','B'),
                    ('G','F')]
```

After creating a list of edges, this list has to be parallelized by using the `parallelize()` function. We are parallelizing this data into four partitions:

```
>>> edgeRDD = sc.parallelize(edgeDataList, 4)
```

We have created an RDD of the edges list.

```
>>> edgeRDD.take(4)
```

Here is the output:

```
[('A', 'C'),
 ('A', 'B'),
 ('B', 'A'),
 ('B', 'C')]
```

After creating the RDD of edges, the RDD of rows has to be created to create the DataFrame. The following line of code creates an RDD of Row objects:

```
>>> edgeRDDRows = edgeRDD.map( lambda data : Row(data[0], data[1]))
>>> edgeRDDRows.take(4)
```

Here is the output:

```
[<Row(A, C)>,
 <Row(A, B)>,
 <Row(B, A)>,
 <Row(B, C)>]
```

A schema is required for our edge DataFrame. We have to create a column schema for the source node column and destination node column. Then, using the `StructType()` function, we will create a schema for our edge DataFrame.

```
>>> sourceColumn = StructField('src', StringType(),True)
>>> destinationColumn = StructField('dst', StringType(), True)
>>> edgeSchema = StructType([sourceColumn, destinationColumn])
```

Have you observed that for sourceColumn we have given the name as src, and for destinationColumn, we have given the name dst? This is also mandatory; it is required syntax for GraphFrames. The schema for the DataFrame has been created. The next step, obviously, is to create the DataFrame:

```
>>> edgeDataFrame = sqlContext.createDataFrame(edgeRDDRows, edgeSchema)

>>> edgeDataFrame.show(5)
```

Here is the output, showing only the top five rows:

```
+---+---+
|src|dst|
+---+---+
|  A|  C|
|  A|  B|
|  B|  A|
|  B|  C|
|  B|  G|
+---+---+
```

Step 8-6-3. Creating a GraphFrames Object

At this moment, we have verticesDataFrame, a DataFrame of vertices; and edgeDataFrame, a DataFrames of edges. Using these two, we can create our graph. In GraphFrames, we can create a graph by using the following code lines:

```
>>> import graphframes.graphframe  as gfm
>>> ourGraph = gfm.GraphFrame(verticesDataFrame, edgeDataFrame)
```

The GraphFrame Python class is defined under the graphframes.graphframe submodule. GraphFrame() takes the vertices and edges DataFrames and returns a GraphFrames object. We have our GraphFrames object, ourGraph. We can fetch all the vertices as follows:

```
>>> ourGraph.vertices.show(5)
```

Here is the output, showing only the top five rows:

```
+---+
| id|
+---+
|  A|
|  B|
|  C|
|  D|
|  E|
+---+
```

We can fetch edges also; here are the top five rows:

```
>>> ourGraph.edges.show(5)
```

```
+---+---+
|src|dst|
+---+---+
|  A|  C|
|  A|  B|
|  B|  A|
|  B|  C|
|  B|  G|
+---+---+
```

Step 8-6-4. Running a Breath-First Search Algorithm

We have created a graph from the required data. Now we can run a breadth-first algorithm on the ourGraph graph. The bfs() function is a breadth-first search (BFS) implementation in GraphFrames. This function is defined on the GraphFrames object. Therefore, we can run this BFS on our GraphFrames object ourGraph. We want to get the minimum path between node D and node G. The first argument, fromExpr, is an expression that tells us that we have to start our BFS from node D. The second argument is toExpr, and the value of toExpr indicates that the destination node for our search is G.

```
>>> bfsPath = ourGraph.bfs(fromExpr="id='D'", toExpr = "id='G'")
>>> bfsPath.show()
```

Here is the output:

```
+----+-----+---+-----+---+
|from|   e0| v1|   e1| to|
+----+-----+---+-----+---+
| [D]|[D,F]|[F]|[F,G]|[G]|
+----+-----+---+-----+---+
```

The output of BFS is very clear. The [D] in the from column means that the start node for BFS is D. The [G] in the to column indicates that the destination node of BFS is G. The shortest path between D and G is from D to F and from F to G.

■ **Note** You can read more about GraphFrames in the following web pages:

https://graphframes.github.io/user-guide.html

https://graphframes.github.io/quick-start.html

https://graphframes.github.io/api/python/index.html

Recipe 8-7. Calculate Page Rank Using GraphFrames

Problem

You want to perform a page-rank algorithm using GraphFrames.

Solution

We already discussed page rank in Chapter 5. We have been given the same graph of web pages that we used in the previous chapter. You want to run the page-rank algorithm using DataFrames. Figure 8-2 shows the network of web pages.

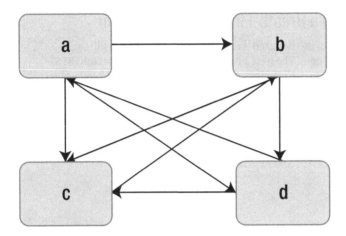

Figure 8-2. *Graph of web pages*

How It Works

First, we have to create a GraphFrames object for the given graph. We have to then create a DataFrame of vertices and a DataFrame of edges.

Step 8-7-1. Creating DataFrame of Vertices

We have been given a graph of four nodes. Our four nodes are a, b, c, and d. First, we create a list of these four nodes:

```
>>> verticesList  = ['a', 'b', 'c', 'd']
>>> verticesListRDD = sc.parallelize(verticesList, 4)
>>> verticesListRowsRDD = verticesListRDD.map( lambda data : Row(data))
>>> verticesListRowsRDD.collect()
```

Here is the output:

```
[<Row(a)>,
<Row(b)>,
<Row(c)>,
<Row(d)>]
```

```
>>> verticesSchema = StructType([StructField('id', StringType(), True)])
>>> verticesDataFrame = sqlContext.createDataFrame(verticesListRowsRDD,
verticesSchema)
>>> verticesDataFrame.show()
```

```
+---+
| id|
+---+
|  a|
|  b|
|  c|
|  d|
+---+
```

We have created a DataFrame of vertices. Now we have to create a Data Frame of edges.

Step 8-7-2. Creating a DataFrame of Edges

To create a DataFrame of edges, the steps are similar to those in many previous recipes. First, we have to create a list of edges. Each edge in the list will be defined by a tuple. Then, we have to create an RDD of edges. Thereafter, we have to transform our RDD to an RDD of row objects. This will be followed by creating a schema and a DataFrame of edges. Let's perform the steps:

```
>>> edgeDataList = [('a','b'), ('a','c'), ('a','d'), ('b', 'c'),
                    ('b', 'd'),('c', 'b'), ('d', 'a'), ('d', 'c')]
>>> sourceColumn = StructField('src', StringType(),True)
>>> destinationColumn = StructField('dst', StringType(), True)
>>> edgeSchema = StructType([sourceColumn, destinationColumn])
>>> edgeRDD = sc.parallelize(edgeDataList, 4)
>>> edgeRDD.take(4)
```

Here is the output:

```
[('a', 'b'),
 ('a', 'c'),
 ('a', 'd'),
 ('b', 'c')]
```

```
>>> edgeRDDRows = edgeRDD.map( lambda data : Row(data[0], data[1]))
>>> edgeRDDRows.take(4)
```

Here is the output:

```
[<Row(a, b)>,
<Row(a, c)>,
<Row(a, d)>,
<Row(b, c)>]
```

```
>>> edgeDataFrame = sqlContext.createDataFrame(edgeRDDRows, edgeSchema)
>>> edgeDataFrame.show(5)
```

Here is the output, showing only the top five rows:

```
+---+---+
|src|dst|
+---+---+
|  a|  b|
|  a|  c|
|  a|  d|
|  b|  c|
|  b|  d|
+---+---+
```

We have created a DataFrame of edges. Let's create a graph.

Step 8-7-3. Creating a Graph

The process of creating a graph follows the same path as in the preceding recipe:

```
>>> import graphframes.graphframe  as gfm
>>> ourGraph = gfm.GraphFrame(verticesDataFrame, edgeDataFrame)
>>> ourGraph.vertices.show(5)
```

Here is the output:

```
+---+
| id|
+---+
|  a|
|  b|
|  c|
|  d|
+---+
```

```
>>> ourGraph.edges.show(5)
```

Here is the output, showing only the top five rows:

```
+---+---+
|src|dst|
+---+---+
|  a|  b|
|  a|  c|
|  a|  d|
|  b|  c|
|  b|  d|
+---+---+
```

Step 8-7-4. Running a Page-Rank Algorithm

Page rank for pages can be found by using the pageRank() function, which is defined on the GraphFrames object:

```
>>> pageRanks = ourGraph.pageRank(resetProbability=0.15, tol=0.01)
```

You might be wondering about the return type of the pageRank() function. Let's see by printing it:

```
>>> pageRanks
```

Here is the output:

```
GraphFrame(v:[id: string, pagerank: double], e:[src: string, dst:
string ... 1 more field])
```

The pageRank() function returns a GraphFrame object. The returned GraphFrame object has vertices and edges. The vertices part of the returned GraphFrame object has the web pages and corresponding page ranks. Let's explore what is inside the edges part of pageRanks:

```
>>> pageRanks.edges.show()
```

Here is the output:

```
+---+---+------------------+
|src|dst|            weight|
+---+---+------------------+
|  d|  a|               0.5|
|  b|  d|               0.5|
|  a|  b|0.3333333333333333|
|  a|  d|0.3333333333333333|
|  d|  c|               0.5|
|  a|  c|0.3333333333333333|
|  b|  c|               0.5|
|  c|  b|               1.0|
+---+---+------------------+
```

It can be observed that the edges part of the pageRanks object consists of edges and corresponding weights:

```
>>> pageRanks.vertices.select('id','pagerank')
```

Here is the output:

```
DataFrame[id: string, pagerank: double]
```

```
>>> pageRanks.vertices.select('id','pagerank').show()
```

Here is the output:

```
+---+------------------+
| id|          pagerank|
+---+------------------+
|  a|0.4831888601952005|
|  b| 1.238562817904233|
|  d| 0.806940642367432|
|  c|1.1401295025626326|
+---+------------------+
```

Finally, we have the page rank for the given pages.

Recipe 8-8. Read Data from Apache Hive
Problem
You want to read table data from Apache Hive.

Solution
We have a table, filamentdata, in Hive. This is the same filament data we have used in many recipes. We have to read this data by using PySparkSQL from Apache Hive. Let's look at the whole process. First we are going to create a table in Hive and upload data into it. Let's start with creating the table filamentdata. We'll create our table in the apress database of Hive. We created this database in Chapter 2, at the time of installation. But let's check that our creation still exists. We can display all the databases in Hive by using show:

```
hive> show databases;
```

Here is the output:

```
OK
apress
default
Time taken: 3.275 seconds, Fetched: 2 row(s)
```

We have the database apress. Therefore, we have to use this database by using the use command:

```
hive> use apress;
```

Here is the output:

```
OK
Time taken: 0.125 seconds
```

After using the database, we create a table named filamenttable by using the following command:

```
hive> create table filamenttable (
    >   filamenttype string,
    >   bulbpower string,
    >    lifeinhours float
    >)
    > row format delimited
    > fields terminated by ',';
```

We have created a a Hive table with three columns. The first column is filamenttype, with values of the string type. The second column is bulbpower, with the data type string. The third column is lifeinhours, of the float type. And now we can display our table by using the show command:

```
hive> show tables;
```

Here is the output:

```
OK
filamenttable
Time taken: 0.118 seconds, Fetched: 1 row(s)
```

The required table has been created successfully. Let's load the data into the table we have created. We'll load the data into Hive from a local directory by using load. The local clause in the following command tells Hive that the data is being loaded from a file in a local directory, not from HDFS.

```
hive> load data local inpath '/home/pysparkbook/pysparkBookData/
filamentData.csv' overwrite into table filamenttable;
```

Here is the output:

```
Loading data to table apress.filamenttable
OK
Time taken: 5.39 seconds
```

After the data loads, we can query the table. We can display a row by using `select` with `limit` to limit the number of rows:

```
hive> select * from filamenttable limit 5;
OK
filamentA        100W      605.0
filamentB        100W      683.0
filamentB        100W      691.0
filamentB        200W      561.0
filamentA        200W      530.0
Time

taken: 0.532 seconds, Fetched: 5 row(s)
```

We have displayed some of the rows of our `filamenttable` table. We have to read this table data by using PySparkSQL.

How It Works
Step 8-8-1. Creating a HiveContext Object

The HiveContext class has been defined inside the `pyspark.sql` submodule. We can create the HiveContext object by using this class and providing `sc` as input to the HiveContext constructor:

```
>>> from pyspark.sql import HiveContext
>>> ourHiveContext = HiveContext(sc)
```

We have created the HiveContext object.

Step 8-8-2. Reading Table Data from Hive

We can read the table by using the `table()` function, which is defined on the HiveContext object. In the `table()` function, we have to provide the name of the table in the format `<databaseName>.<tableName>`. In our case, the database name is `apress`, and the table name is `filamenttable`. Therefore, the argument value for the `table()` function will be `apress.filamenttable`.

```
>>> FilamentDataFrame = ourHiveContext.table('apress.filamenttable')
>>> FilamentDataFrame.show(5)
+------------+---------+-----------+
|filamenttype|bulbpower|lifeinhours|
+------------+---------+-----------+
|    filamentA|     100W|      605.0|
|    filamentB|     100W|      683.0|
|    filamentB|     100W|      691.0|
|    filamentB|     200W|      561.0|
|    filamentA|     200W|      530.0|
+------------+---------+-----------+
only showing top 5 rows
```

And, finally, we have created a DataFrame from the table in Apache Hive.

■ ■ ■

PySpark MLlib and Linear Regression

Machine learning has gone through many recent developments and is becoming more popular day by day. People from all domains, including computer science, mathematics, and management, are using machine learning in various projects to find hidden information in data. Big data becomes more interesting when we start applying machine-learning algorithms to it.

PySpark MLlib is a machine-learning library. It is a wrapper over PySpark Core to do data analysis using machine-learning algorithms. It works on distributed systems and is scalable. We can find implementations of classification, clustering, linear regression, and other machine-learning algorithms in PySpark MLlib. We know that PySpark is good for iterative algorithms. Using iterative algorithms, many machine-learning algorithms have been implemented in PySpark MLlib. Apart from PySpark efficiency and scalability, PySpark MLlib APIs are very user-friendly.

Software libraries, which are defined to provide solutions for various problems, come with their own data structures. These data structures are provided to solve a specific set of problems with efficient options. PySpark MLlib comes with many data structures, including dense vectors, sparse vectors, and a local and distributed matrix.

Linear regression is one of the most popular machine-learning algorithms. We create a linear mathematical model between one or more independent variables and one dependent variable.

In this chapter, we will first move through the various data structures defined in PySpark MLlib. Then we will explore linear regression with MLlib. We will also implement linear regression by using the ridge and lasso methods.

This chapter covers the following recipes:

Recipe 9-1. Create a dense vector

Recipe 9-2. Create a sparse vector

Recipe 9-3. Create local matrices

Recipe 9-4. Create a RowMatrix

Recipe 9-5. Create a labeled point

Recipe 9-6. Apply linear regression

© Raju Kumar Mishra 2018
R. K. Mishra, *PySpark Recipes*, https://doi.org/10.1007/978-1-4842-3141-8_9

Recipe 9-1. Create a Dense Vector
Problem

You want to create a dense vector.

Solution

DenseVector is a local vector. In PySpark, we can create a dense vector by using the DenseVector constructor.

How It Works

Creating a dense vector requires us to import the DenseVector class, which has been defined inside the pyspark.mllib.linalg submodule:

```
>>> from pyspark.mllib.linalg import DenseVector
>>> denseDataList = [1.0,3.4,4.5,3.2]
```

We have created a list of four elements. This list, denseDataList, has floating-point data elements. We have already imported DenseVector, so we are going to create a dense vector now:

```
>>> denseDataVector = DenseVector(denseDataList)
>>> print denseDataVector
```

Here is the output:

```
[1.0,3.4,4.5,3.2]
```

We have created a dense vector named denseDataVector. We can index our DenseVector elements. The elements of DenseVector are zero-indexed. In the following lines of code, we'll first fetch the second element of DenseVector, and then we'll fetch the first element with index 0. Last, we'll fetch the third element of denseDataVector by using the index 2.

```
>>> denseDataVector[1]
```

Here is the output:

```
3.3999999999999999
```

```
>>> denseDataVector[0]
```

Here is the output:

```
1.0
```

```
>>> denseDataVector[2]
```

Here is the output:

```
4.5
```

■ **Note**   We have already discussed NumPy. It is recommended to use a NumPy array over a Python list to create a dense vector, for efficiency. You can read more about DenseVector and SparseVector at https://spark.apache.org/docs/2.0.0-preview/mllib-data-types.html.

Recipe 9-2. Create a Sparse Vector
Problem

You want to create a sparse vector.

Solution

Sometimes we get a dataset in which the maximum value of data elements is 0. Can we escape save zero? Yeah, we can do it. We can save a sparse vector as a SparseVector object in PySpark MLlib.

You have been given a vector. Many elements of this vector have a value of 0, as you can see in Table 9-1. The value at indices 1, 2, 3, 4, 5, and 7 is 0.0.

Table 9-1. *A Sparse Dataset*

Data	1.0	0.0	0.0	0.0	0.0	0.0	3.2	0.0
Index	0	1	2	3	4	5	6	7

So six out of eight values are 0. We are going to create a SparseVector out of this.

How It Works

In this section, we are going to create a SparseVector. First, we have to import the SparseVector class from our submodule pyspark.mllib.linalg:

```
>>> from pyspark.mllib.linalg import SparseVector
>>> sparseDataList = [1.0,3.2]
>>> sparseDataVector = SparseVector(8,[0,7],sparseDataList)
```

We have created a sparse vector named sparseDataVector. Let me explain the arguments of SparseVector. The first argument, 8, indicates that we are have eight elements in our SparseVector. The second element is a list, a list of indices that includes a nonzero element in our SparseVector. We have 1.0 at index 0, and 3.2 at index 7.

Let's print our sparseDataVector and see the result:

```
>>> sparseDataVector
```

Here is the output:

```
SparseVector(8, {0: 1.0, 7: 3.2})
```

You can see that it has our numbers at index 0 and index 7. We can also index our SparseVector by using [ ]:

```
>>> sparseDataVector[1]
```

Here is the output:

```
0.0
```

```
>>> sparseDataVector[7]
```

Here is the output:

```
3.2000000000000002
```

So we have found that at index 1, we have 0.0; and at index 7, we have 3.2. The total count of nonzero elements can be fetched by using the function numNonzeros(). In the following code line, we have found that in our SparseVector, we have only two nonzero elements:

```
>>> sparseDataVector.numNonzeros()
```

Here is the output:

```
squared_distance( )
```

```
2
```

We can do many operations on a SparseVector. The following code line shows how to calculate the squared distance between two given SparseVectors. We calculate the squared distance by using the squared_distance() function:

```
>>> sparseDataList1 = [3.0,1.4,2.5,1.2]
>>> sparseDataVector1 = SparseVector(8,[0,3,4,6],sparseDataList1)
>>> squaredDistance = sparseDataVector.squared_distance(sparseDataVector1)
>>> squaredDistance
```

Here is the output:

```
23.890000000000001
```

We have the squared distance.

Recipe 9-3. Create Local Matrices
Problem
You want to create local matrices.

Solution

A local matrix is stored on a single machine, and, obviously, it is local to that machine. The indices of a local matrix are of the integer type. And the values of a local matrix are of the double type. A local matrix comes in two flavors: a dense matrix and a sparse matrix. The elements of a dense matrix are stored in a single array, in column major order. In a sparse matrix, nonzero values are stored in a compressed sparse column format. We can create a dense matrix by using the dense() method defined in the Matrices class. The dense() method is a static method, so it is not necessary to create an object of the Matrices class. Similarly, we can create a sparse local matrix by using the sparse() method defined in the Matrices class. This method is also a static method.

How It Works

First, we will create a local dense matrix. Then we will create a local sparse matrix. Let's first create a Python list:

```
>>> denseDataList = [1.0,3.4,4.5,3.2]

>>> ourDenseMatrix = localMtrix.dense(numRows = 2, numCols = 2, values=
denseDataList)
```

We have created a local dense matrix, ourDenseMatrix. We should discuss the arguments of the dense() method. We have provided three arguments to it. The first argument defines the number of rows in our matrix. The second argument is the number of columns. The third argument is a list of matrix elements. Now let's print the matrix we have created and see how it looks:

```
>>> ourDenseMatrix
```

Here is the output:

```
DenseMatrix(2, 2, [1.0, 3.4, 4.5, 3.2], False)
```

A new thing appears—False—in the result. What is this? This tells us whether the matrix is transposed.

You might be wondering how to visualize the structure of the dense matrix that we have created. The function to help us is toArray(). This will transform our dense matrix to a NumPy array:

```
>>> ourDenseMatrix.toArray()
```

Here is the output:

```
array([[ 1. ,  4.5],
       [ 3.4,  3.2]])
```

We can see that our data points have been filled into the dense matrix by column. Let me explain The first two values of denseDataList list are filled in first in the dense matrix, in the first column. Then the remaining two elements are placed in the second column.

We create a sparse matrix with the following line of code:

```
>>> sparseDataList = [1.0,3.2]
```

Our sparse matrix has only two nonzero elements. We want to create a matrix of 2 × 2. Therefore, it is obvious that our matrix also will have two 0 elements. We know that PySpark saves Spark local matrices in a compressed sparse column format. Therefore, we have to provide column pointers as an argument. We want to create a diagonal matrix using the numbers 1.0 and 3.2:

```
>>> ourSparseMatrix = localMtrix.sparse(numRows = 2,  numCols = 2, colPtrs =
[0,1,2],  rowIndices = [0,1], values = sparseDataList)
>>> ourSparseMatrix.toArray()
```

Here is the output:

```
array([[ 1. ,  0. ],
       [ 0. ,  3.2]])
```

■ **Note**   You can read more about compressed sparse column format on Wikipedia, https://en.wikipedia.org/wiki/Sparse_matrix.

Recipe 9-4. Create a Row Matrix
Problem

You want to create a row matrix.

Solution

RowMatrix is a distributed matrix. The row indices of RowMatrix are meaningless. It is a row-oriented distributed matrix.

How It Works

The class RowMatrix is in the PySpark submodule pyspark.mllib.linalg.distributed. Let's first import this class. Then we will create a RowMatrix.

```
>>> from pyspark.mllib.linalg.distributed import RowMatrix as rm
```

To create a RowMatrix, we first need an RDD of vectors. But even if you have an RDD of lists, that will work. We have our data in a nested list, dataList:

```
>>> dataList = [[ 94.88,   82.04,   52.57],
...             [ 35.85,   26.9 ,    3.63],
...             [ 41.76,   69.67,   50.62],
...             [ 90.45,   54.66,   64.07]]
```

The nested list dataList has four rows and three columns. Now we have to create an RDD out of this dataList:

```
>>> dataListRDD = sc.parallelize(dataList,4)
>>> ourRowMatrix = rm(rows = dataListRDD, numRows = 4 , numCols = 3)
```

We have created our RowMatrix. The first argument of RowMatrix is the RDD of the list. The second and third arguments are the number of rows and the number of columns, respectively.

```
>>> ourRowMatrix.numRows()
```

Here is the output:

```
4L
```

```
>>> ourRowMatrix.numCols()
```

Here is the output:

3L

Recipe 9-5. Create a Labeled Point
Problem
You want to create a labeled point.

Solution

A *labeled point* is a basic data structure for linear regression and classification algorithms. It consists of a sparse or dense vector associated with a label. Labels are of the double data type; therefore, a labeled point can be used in both regression and classification. The LabeledPoint class stays inside the pyspark.mllib.regression submodule. Let's create LabeledPoint data.

How It Works

First we define a nested list:

```
>>> from pyspark.mllib.regression import LabeledPoint
>>> labeledPointData = [[3.09,1.97,3.73,1],
...                     [2.96,2.15,4.16,1],
...                     [2.87,1.93,4.39,1],
...                     [3.02,1.55,4.43,1],
...                     [1.8,3.65,2.08,2],
...                     [1.36,4.43,1.95,2],
...                     [1.71,4.35,1.94,2],
...                     [1.03,3.75,2.12,2],
...                     [2.3,3.59,1.99,2]]
```

The last data point in every list of our nested list labeledPointData is our label. Let's parallelize the data:

```
>>> labeledPointDataRDD = sc.parallelize(labeledPointData, 4)
>>> labeledPointDataRDD.take(4)
```

Here is the output:

```
[[3.09, 1.97, 3.73, 1],
 [2.96, 2.15, 4.16, 1],
 [2.87, 1.93, 4.39, 1],
 [3.02, 1.55, 4.43, 1]]
```

```
>>> labeledPointRDD = labeledPointDataRDD.map(lambda data : LabeledPoint
(data[3],data[0:3]))
>>> labeledPointRDD.take(4)
```

Here is the output:

```
[LabeledPoint(1.0, [3.09,1.97,3.73]),
 LabeledPoint(1.0, [2.96,2.15,4.16]),
 LabeledPoint(1.0, [2.87,1.93,4.39]),
 LabeledPoint(1.0, [3.02,1.55,4.43])]
```

We have created our labeledPointRDD. You can see that the labeled points have been transformed into double data types.

```
https://spark.apache.org/docs/2.0.0/mllib-data-types.html
```

Recipe 9-6. Apply Linear Regression
Problem

You want to apply linear regression.

Solution

Linear regression is a supervised machine-learning algorithm. Here we fit a line (a straight line or a curved line) that generates a linear relationship between dependent and independent variables.

We have been given a file, linearRegressionData.csv. This file consists of four columns. If we visualize the file data, it looks as follows:

```
+-----+----+----+----+
| dvs1|ivs1|ivs2|ivs3|
+-----+----+----+----+
|34.63|5.53|5.58|5.41|
|40.89|3.89|6.48|6.97|
|37.25|5.07| 4.5| 6.5|
|45.09|5.81|5.71|8.59|
| 39.4|5.61|5.79|6.77|
+-----+----+----+----+.
```

The column dvs1 is the dependent variable, which depends on the three independent variables ivs1, ivs2, and ivs3. We have to create a mathematical model that will show a linear relationship between the dependent and independent variables. The linear regression model's mathematical formula is depicted in Figure 9-1.

$$dvs\,1 = \beta_0 + \beta_1\,ivs\,1 + \beta_2\,ivs\,2 + \beta_3\,ivs\,3$$
$$Where\ \beta_0\ is\ Intercept$$
$$\beta_1,\ \beta_2\ and\ \beta_3\ are\ coefficients\ or\ weights$$

Figure 9-1. *Mathematical formula for the linear regression model*

We have to estimate the value of the intercept and weights.

How It Works

Let's create the model step-by-step.

Step 9-6-1. Reading CSV File Data

Our regression data is in `linearRegressionData.csv`. We have to read this data and transform it into labeled points in order to run linear regression analysis on it. I think that a better strategy is to transform our data to an RDD of labeled points that can be accomplished by first reading the file by using the `spark.read.csv()` function. We know that the `spark.read.csv()` function will return a DataFrame. We have to transform that DataFrame to an RDD of labeled points somehow. So first let's read the file.

```
>>> regressionDataFrame = spark.read.csv('file:///home/pysparkbook/bData/
linearRegressionData.csv',header=True, inferSchema = True)
>>> regressionDataFrame.show(5)
```

Here is the output, showing only the top five rows:

```
+-----+----+----+----+
| dvs1|ivs1|ivs2|ivs3|
+-----+----+----+----+
|34.63|5.53|5.58|5.41|
|40.89|3.89|6.48|6.97|
|37.25|5.07| 4.5| 6.5|
|45.09|5.81|5.71|8.59|
| 39.4|5.61|5.79|6.77|
+-----+----+----+----+.
```

After reading the file, we have the DataFrame. The following line of code transforms our DataFrame to an RDD:

```
>>> regressionDataRDDDict = regressionDataFrame.rdd
>>> regressionDataRDDDict.take(5)
```

Here is the output:

```
[Row(dvs1=34.63, ivs1=5.53, ivs2=5.58, ivs3=5.41),
 Row(dvs1=40.89, ivs1=3.89, ivs2=6.48, ivs3=6.97),
 Row(dvs1=37.25, ivs1=5.07, ivs2=4.5, ivs3=6.5),
 Row(dvs1=45.09, ivs1=5.81, ivs2=5.71, ivs3=8.59),
 Row(dvs1=39.4, ivs1=5.61, ivs2=5.79, ivs3=6.77)]
```

We have seen the output of regressionDataRDDDict after transformation into an RDD. This is an RDD of row objects. You might be thinking that you know about row objects. We use these while creating a DataFrame. But this is not our requirement. You can see that the row objects have data in key/value format. So we need more transformations to get only the values in our RDD:

```
>>> regressionDataRDD = regressionDataFrame.rdd.map(list)
>>> regressionDataRDD.take(5)
```

Here is the output:

```
[[34.63, 5.53, 5.58, 5.41],
 [40.89, 3.89, 6.48, 6.97],
 [37.25, 5.07, 4.5, 6.5],
 [45.09, 5.81, 5.71, 8.59],
 [39.4, 5.61, 5.79, 6.77]]
```

Adding a list as an argument of the RDD map() function has transformed our data into a format that we can use to easily create our labeled point RDD. In the following step, we are going to create a LabeledPoint RDD.

Step 9-6-2. Creating an RDD of the Labeled Point

In order to run linear regression, we have transformed our data into labeled points. As we discussed, the first column of our RDD is a dependent variable, which depends on the rest of the variables. Therefore, the first value of every RDD element is our label, and the rest are our features. Now we can create the LabeledPoint RDD. We know that to use LabeledPoint, we have to import the class:

```
>>> from pyspark.mllib.regression import LabeledPoint
>>> regressionDataLabelPoint = regressionDataRDD.map(lambda data : LabeledPoint(data[0],data[1:4]))
```

The map() function can be used to transform our RDD to a LabeledPoint RDD. The first argument of LabeledPoint is the label, which we provide as data[0] in the preceding code. We are going to take five elements out of the LabeledPoint RDD via regressionDataLabelPoint. Surely, we are going to use the take() function with 5 as an argument to it:

```
>>> regressionDataLabelPoint.take(5)
```

Here is the output:

```
[LabeledPoint(34.63, [5.53,5.58,5.41]),
 LabeledPoint(40.89, [3.89,6.48,6.97]),
 LabeledPoint(37.25, [5.07,4.5,6.5]),
 LabeledPoint(45.09, [5.81,5.71,8.59]),
 LabeledPoint(39.4, [5.61,5.79,6.77])]
```

We have created the required LabeledPoint RDD. As we discussed, linear regression is a supervised machine-learning algorithm. Therefore, we first divide our given datasets into training and testing datasets. The training dataset will be used to create the model, and then we'll apply the testing data to check the accuracy of the linear regression model we have created. In the following step, we will divide our dataset into training and testing datasets.

Step 9-6-3. Dividing Training and Testing Data

PySpark provides the randomSplit() function, which we can use to divide our datasets into training and testing datasets:

```
>>> regressionLabelPointSplit = regressionDataLabelPoint.
randomSplit([0.7,0.3])
```

We providing the list [0.7 , 0.3] as an argument. This list indicates that we need 70 percent of our data points in our training dataset, and the rest in our testing dataset. In our dataset, we have a total of 30 records. Therefore, 22 records will go in the training dataset, and 8 will go in the testing dataset.

```
>>> regressionLabelPointTrainData = regressionLabelPointSplit[0]
>>> regressionLabelPointTrainData.take(5)
```

Here is the output:

```
[LabeledPoint(34.63, [5.53,5.58,5.41]),
 LabeledPoint(45.09, [5.81,5.71,8.59]),
 LabeledPoint(39.4, [5.61,5.79,6.77]),
 LabeledPoint(33.25, [5.33,5.78,4.94]),
 LabeledPoint(44.17, [6.11,6.18,8.2])]
```

The training dataset is ready:

```
>>> regressionLabelPointTrainData.count()
```

Here is the output:

```
22
```

As we discussed, there are 22 data points are in our training dataset:

```
>>> regressionLabelPointTestData = regressionLabelPointSplit[1]
>>> regressionLabelPointTestData.take(5)
```

Here is the output:

```
[LabeledPoint(40.89, [3.89,6.48,6.97]),
 LabeledPoint(37.25, [5.07,4.5,6.5]),
 LabeledPoint(42.92, [5.39,6.59,7.56]),
 LabeledPoint(38.23, [4.19,6.47,6.52]),
 LabeledPoint(36.33, [5.65,5.78,5.47])]
```

```
>>> regressionLabelPointTestData.count()
```

Here is the output:

8

The testing dataset has been created as regressionLabelPointTestData. The count() function on the test data ensures that we have eight data points in our testing dataset.

Now that we have the training and testing datasets, we are ready to fit the regression model to our training dataset. In the following step, we are going to create our linear regression model.

Step 9-6-4. Creating a Linear Regression Model

PySpark uses *stochastic gradient descent* (SGD) to calculate the coefficients of the linear regression model. This is an optimization algorithm. The PySpark class LinearRegressionWithSGD is used to do the operation related to linear regression.

Let's import LinearRegressionWithSGD; we'll import it as lrSGD:

```
>>> from pyspark.mllib.regression import LinearRegressionWithSGD as lrSGD
>>> ourModelWithLinearRegression  = lrSGD.train(
                                    data = regressionLabelPointTrainData,
...                                        iterations = 200,
...                                        step = 0.02,
...                                        intercept = True)
```

There is a static method, train(), defined in the LinearRegressionWithSGD class. The train() method is used to create a linear regression model. The first argument of the train() method is the data that is our training data. SGD is an iterative algorithm, so we provide the number of iterations as the second argument to our train() method. The third parameter, step, defines the size of the movement in the SGD algorithm. If a linear model has an intercept, we have to set that intercept to True. Finally, this creates our linear regression model named ourModelWithLinearRegression.

```
>>> ourModelWithLinearRegression.intercept
```

Here is the output:

```
1.3475409224629387
```

```
>>> ourModelWithLinearRegression.weights
```

Here is the output:

```
DenseVector([1.7083, 2.1529, 2.3226])
```

We can create our regression model by using the intercept and weights. The created model is shown in in Figure 9-2.

$$dvs\,1 = 1.34 + 1.7\,ivs\,1 + 2.15\,ivs\,2 + 2.32\,ivs\,3$$

Figure 9-2. *Our regression model*

■ **Note** You can read more about stochastic gradient descent on Wikipedia, https://en.wikipedia.org/wiki/Stochastic_gradient_descent.

Step 9-6-5. Saving the Created Model

Sometimes we have to save the created model and use it in the future. We can save our model by using the save() method:

```
>>> ourModelWithLinearRegression.save(sc, '/home/pysparkbook/
ourModelWithLinearRegression')
```

The first argument of the save() method is SparkContext. The second argument is the path of the directory where you want to save your model. We have saved our model, but how do we read it? Reading the saved model requires the load() method. This method is inside the LinearRegressionModel class.

```
>>> from pyspark.mllib.regression import LinearRegressionModel as
linearRegressModel
```

```
>>> ourModelWithLinearRegressionReloaded = linearRegressModel.load(sc, '/
home/pysparkbook/ourModelWithLinearRegression')
```

The saved model has been reloaded, so now we can use it. Since we have reloaded the model, let's check whether we can get the intercept and weight values from the reloaded model:

```
>>> ourModelWithLinearRegressionReloaded.intercept
```

Here is the output:

```
1.34754092246
>>> ourModelWithLinearRegressionReloaded.weights
```

Here is the output:

```
DenseVector([1.7083, 2.1529, 2.3226])
```

It is clear that we have reloaded the model.

Step 9-6-6. Predicting Data by Using the Model

Whenever any mathematical model is created, we can check the credibility of the model. To check our model's credibility, we will make an RDD of the actual and predicted data of our test data:

```
>>> actualDataandLinearRegressionPredictedData = regressionLabelPointTest
Data.map(lambda data : (float(data.label) , float(ourModelWithLinearRegressi
on.predict(data.features))))
```

The predict() method will take the features data (which are independent variables) and return the predicted values for the dependent variable. Now we'll have an RDD of the actual and predicted data.

```
>>> actualDataandLinearRegressionPredictedData.take(5)
```

Here is the output:

```
[(40.89, 38.1322613341641),
 (37.25, 34.79375295252528),
 (42.92, 42.30191258048799),
 (38.23, 37.57804867136557),
 (36.33, 36.14796052442989)]
```

Step 9-6-7. Evaluating the Model We Created

After getting the RDD of actual and predicted data, we will calculate evaluation metrics. We will first calculate the root-mean-square error. The mathematical formula for a root-mean-square error is given in Figure 9-3.

dvs1 is the actual value of the dependent variable and dvs1pred is the predicted value of the dependent variable. We have n data points.

$$\text{Root-mean-squared Error} = \sqrt{\frac{\sum (dvs1 - dvs1\,pred)^2}{n}}$$

Figure 9-3. *Mathematical formula for a root-mean-square error*

We have to import the RegressionMetrics class to evaluate the model:

```
>>> from pyspark.mllib.evaluation import RegressionMetrics as rmtrcs
>>> ourLinearRegressionModelMetrics = rmtrcs(actualDataandLinearRegressionP
redictedData)
>>> ourLinearRegressionModelMetrics.rootMeanSquaredError
```

Here is the output:

```
1.8446573587605941
```

The value of our root-mean-square error is 1.844657. Similarly, we calculate the value of R^2:

```
>>> ourLinearRegressionModelMetrics.r2
```

Here is the output:

```
0.47423120771913974
```

The value of R^2 is less, even less than 0.5. So is this a good model? I say no. So another question is, can we improve the efficiency of our model? And the answer is yes; I have done it just by playing with the learning step size and number of iterations:

```
>>> ourModelWithLinearRegression  = lrSGD.train(data = regressionLabelPoint
                                    TrainData,
...                                                 iterations = 100,
...                                                 step = 0.05,
...                                                 intercept = True)

>>> actualDataandLinearRegressionPredictedData =
regressionLabelPointTestData.map(lambda data : (float(data.label) ,
float(ourModelWithLinearRegression.predict(data.features))))
```

```
>>> from pyspark.mllib.evaluation import RegressionMetrics as rmtrcs

>>> ourLinearRegressionModelMetrics = rmtrcs(actualDataandLinearRegressionP
redictedData)

>>> ourLinearRegressionModelMetrics.rootMeanSquaredError
```

Here is the output:

```
1.7856232547826518
```

```
>>> ourLinearRegressionModelMetrics.r2
```

Here is the output:

```
0.6377723547885376
```

So, finally, we have increased the value of R^2.

■ **Note**   You can read more about R^2 at https://en.wikipedia.org/wiki/
Coefficient_of_determination.

Recipe 9-7. Apply Ridge Regression
Problem

You want to apply ridge regression.

Solution

You have been given a dataset in the CSV file autoMPGDataModified.csv. This dataset
has five columns. We have to fit a linear regression model to this data by using ridge
regularization. The first column is miles per gallon, which is the dependent variable in
this case.

```
+----+------------+----------+------+------------+
| mpg|displacement|horsepower|weight|acceleration|
+----+------------+----------+------+------------+
|18.0|       307.0|        18|  3504|        12.0|
|15.0|       350.0|        36|  3693|        11.5|
|18.0|       318.0|        30|  3436|        11.0|
|16.0|       304.0|        30|  3433|        12.0|
|17.0|       302.0|        25|  3449|        10.5|
+----+------------+----------+------+------------+
```

I have taken this dataset from the UCI Machine Learning Repository (https://archive.ics.uci.edu/ml/datasets/auto+mpg) and removed some columns. According to the web page, the dataset was taken from the StatLib library maintained at Carnegie Mellon University and was used in the 1983 American Statistical Association Exposition.

You might be wondering about the difference between linear regression and linear regression with the ridge parameter. We know that we do optimization of error part using SGD. So in the error part, an extra term is added, as shown in Figure 9-4.

At the time of error minimization we add an extra following expression.

$$\alpha \sum \beta_i^2$$

Where α is regularization parameter

Figure 9-4. *Extra error term in ridge regression*

Let's perform ridge regression on the given dataset.

■ **Note** You can read more about the auto-mpg data on the following sites:

https://archive.ics.uci.edu/ml/datasets/auto+mpg

https://archive.ics.uci.edu/ml/machine-learning-databases/auto-mpg/

You can read more about ridge regression on the following sites:

www.quora.com/Why-is-it-that-the-lasso-unlike-ridge-regression-results-in-coefficient-estimates-that-are-exactly-equal-to-zero

https://en.wikipedia.org/wiki/Tikhonov_regularization

https://en.wikipedia.org/wiki/Regularization_(mathematics)

How It Works
Step 9-7-1. Reading the CSV File Data

We have to read the data and transform it to RDD, as we have done in previous recipes:

```
>>> autoDataFrame = spark.read.csv('file:///home/pysparkbook/bData/
autoMPGDataModified.csv',header=True, inferSchema = True)
>>> autoDataFrame.show(5)
```

Here is the output, showing only the top five rows:

```
+----+------------+----------+------+------------+
| mpg|displacement|horsepower|weight|acceleration|
+----+------------+----------+------+------------+
|18.0|       307.0|        18|  3504|        12.0|
|15.0|       350.0|        36|  3693|        11.5|
|18.0|       318.0|        30|  3436|        11.0|
|16.0|       304.0|        30|  3433|        12.0|
|17.0|       302.0|        25|  3449|        10.5|
+----+------------+----------+------+------------+
```

```
>>> autoDataFrame.printSchema()
```

Here is the output:

```
root
 |-- mpg: double (nullable = true)
 |-- displacement: double (nullable = true)
 |-- horsepower: integer (nullable = true)
 |-- weight: integer (nullable = true)
 |-- acceleration: double (nullable = true)
```

```
>>> autoDataRDDDict = autoDataFrame.rdd
>>> autoDataRDDDict.take(5)
```

Here is the output:

```
[Row(mpg=18.0, displacement=307.0, horsepower=18, weight=3504,
acceleration=12.0),
Row(mpg=15.0, displacement=350.0, horsepower=36, weight=3693,
acceleration=11.5),
Row(mpg=18.0, displacement=318.0, horsepower=30, weight=3436,
acceleration=11.0),
Row(mpg=16.0, displacement=304.0, horsepower=30, weight=3433,
acceleration=12.0),
Row(mpg=17.0, displacement=302.0, horsepower=25, weight=3449,
acceleration=10.5)]
```

We transform our DataFrame to an RDD so that we can transform it further into the LabeledPoint RDD:

```
>>> autoDataRDD = autoDataFrame.rdd.map(list)
>>> autoDataRDD.take(5)
```

Here is the output:

```
[[18.0, 307.0, 18, 3504, 12.0],
 [15.0, 350.0, 36, 3693, 11.5],
 [18.0, 318.0, 30, 3436, 11.0],
 [16.0, 304.0, 30, 3433, 12.0],

 [17.0, 302.0, 25, 3449, 10.5]]
```

Step 9-7-2. Creating an RDD of the Labeled Points

After getting the RDD, we have to transform the RDD to the LabeledPoint RDD:

```
>>> from pyspark.mllib.regression import LabeledPoint
>>> autoDataLabelPoint = autoDataRDD.map(lambda data : LabeledPoint(data[0],
[data[1]/10,data[2],float(data[3])/100,data[4]]))
```

In the dataset, we can see that it is better to normalize the data. Therefore, we divide the displacement by 10 and the weight by 100:

```
>>> autoDataLabelPoint.take(5)
```

Here is the output:

```
[LabeledPoint(18.0, [30.7,18.0,35.04,12.0]),
 LabeledPoint(15.0, [35.0,36.0,36.93,11.5]),
 LabeledPoint(18.0, [31.8,30.0,34.36,11.0]),
 LabeledPoint(16.0, [30.4,30.0,34.33,12.0]),
 LabeledPoint(17.0, [30.2,25.0,34.49,10.5])]
```

Step 9-7-3. Dividing Training and Testing Data

It is time to divide our dataset into training and testing datasets:

```
>>> autoDataLabelPointSplit = autoDataLabelPoint.randomSplit([0.7,0.3])
>>> autoDataLabelPointTrain = autoDataLabelPointSplit[0]
>>> autoDataLabelPointTest = autoDataLabelPointSplit[1]
>>> autoDataLabelPointTrain.take(5)
```

Here is the output:

```
[LabeledPoint(18.0, [30.7,18.0,35.04,12.0]),
 LabeledPoint(15.0, [35.0,36.0,36.93,11.5]),
 LabeledPoint(18.0, [31.8,30.0,34.36,11.0]),
 LabeledPoint(16.0, [30.4,30.0,34.33,12.0]),
 LabeledPoint(17.0, [30.2,25.0,34.49,10.5])]
```

```
>>> autoDataLabelPointTest.take(5)
```

Here is the output:

```
[LabeledPoint(14.0, [45.5,48.0,44.25,10.0]),
 LabeledPoint(15.0, [39.0,41.0,38.5,8.5]),
 LabeledPoint(15.0, [40.0,30.0,37.61,9.5]),
 LabeledPoint(24.0, [11.3,92.0,23.72,15.0]),
 LabeledPoint(26.0, [9.7,51.0,18.35,20.5])]
```

```
>>> autoDataLabelPointTest.count()
```

Here is the output:

```
122
```

```
>>> autoDataLabelPointTrain.count()
```

Here is the output:

```
269
```

Step 9-7-4. Creating a Linear Regression Model

We can create our model by using the `train()` method of the `RidgeRegressionWithSGD` class. Therefore, we first have to import the `RidgeRegressionWithSGD` class and then run the `train()` method:

```
>>> from pyspark.mllib.regression import RidgeRegressionWithSGD  as ridgeSGD
>>> ourModelWithRidge  = ridgeSGD.train(
data = autoDataLabelPointTrain,
iterations = 400,
step = 0.0005,
regParam = 0.05,
intercept = True
)
```

In our `train()` method, there is one more argument, `regParam`, than in our previous recipe. The `regParam` argument is a regularization parameter, alpha, shown previously in Figure 9-4.

```
>>> ourModelWithRidge.intercept
```

Here is the output:

```
1.0192595005891258
```

```
>>> ourModelWithRidge.weights
```

Here is the output:

```
DenseVector([-0.0575, 0.2025, 0.1961, 0.3503])
```

We have created our model and have the intercept and coefficients.

Step 9-7-5. Saving the Created Model

We can save our model and reload it as we did in the previous recipe. The following code first saves the model and then reloads it. After reloading the model, we will check whether it is working correctly.

```
>>> ourModelWithRidge.save(sc, '/home/pysparkbook/ourModelWithRidge')

>>> from  pyspark.mllib.regression import RidgeRegressionModel as
ridgeRegModel

>>> ourModelWithRidgeReloaded = ridgeRegModel.load(sc, '/home/pysparkbook/
ourModelWithRidge')
>>> ourModelWithRidgeReloaded.intercept
```

Here is the output:

```
1.01925950059
```

```
>>> ourModelWithRidgeReloaded.weights
```

Here is the output:

```
DenseVector([-0.0575, 0.2025, 0.1961, 0.3503])
```

Our saved model is working correctly.

Step 9-7-6. Predicting the Data by Using the Model

In this step, we are going to create an RDD of actual and predicted data. The predicted data will be calculated by using the predict() function.

```
>>> actualDataandRidgePredictedData = autoDataLabelPointTest.map(lambda data
: [float(data.label) , float(ourModelWithRidge.predict(data.features))])
>>> actualDataandRidgePredictedData.take(5)
```

Here is the output:

```
[[18.0, 15.857286660271024],
 [16.0, 16.28216643081738],
 [17.0, 14.787196092732607],
 [15.0, 17.60672713589945],
 [14.0, 17.67800889949583]]
```

Step 9-7-7. Evaluating the Model We Have Created

We have to again find the root-mean-square error:

```
>>> ourRidgeModelMetrics = rmtrcs(actualDataandRidgePredictedData)
>>> ourRidgeModelMetrics.rootMeanSquaredError
```

Here is the output:

```
8.149263319131556
```

This is the error value. The higher the value, the less accurate the model is. We have calculated the root-mean-square error, and we have checked the credibility of the model.

Recipe 9-8. Apply Lasso Regression
Problem

You want to apply lasso regression.

Solution

Linear regression with lasso regularization is used for models that are not properly fitted. In the case of lasso, at the time of error minimization, we add the term shown in Figure 9-5.

At the time of error minimization we add an extra following expression.

$$\alpha \sum |\beta_i|$$

Where α is regularization parameter

Figure 9-5. *Extra error term in lasso regression*

How It Works
Step 9-8-1. Creating a Linear Regression Model with Lasso

We have already created LabeledPoint, containing auto data. We can apply the train() method defined in the LassoWithSGD class:

```
>>> from pyspark.mllib.regression import LassoWithSGD  as lassoSGD
>>> ourModelWithLasso  = lassoSGD.train(data = autoDataLabelPointTrain,
iterations = 400, step = 0.0005,regParam = 0.05, intercept = True)
```

We have created our model.

```
>>> ourModelWithLasso.intercept
```

Here is the output:

```
1.020329086499831
```

```
>>> ourModelWithLasso.weights
```

Here is the output:

```
DenseVector([-0.063, 0.2046, 0.198, 0.3719])
```

We have the intercept and weight of the model.

Step 9-8-2. Predicting the Data Using the Lasso Model

In order to get the RDD of actual and predicted data, we are going to use the same strategy used in previous recipes:

```
>>> actualDataandLassoPredictedData = autoDataLabelPointTest.map(lambda data
: (float(data.label) , float(ourModelWithLasso.predict(data.features))))
>>> actualDataandLassoPredictedData.take(5)
```

Here is the output:

```
[(15.0, 17.768596038896607),
 (16.0, 16.5021818747879),
 (17.0, 14.965800201626084),
 (15.0, 17.734571412337576),
 (15.0, 17.154509770352835)]
```

Step 9-8-3. Evaluating the Model We Have Created

Now we have to test the model—and though there's no need to say it, we are going to use the same strategy as before:

```
>>> from pyspark.mllib.evaluation import RegressionMetrics as rmtrcs
>>> ourLassoModelMetrics = rmtrcs(actualDataandLassoPredictedData)
>>> ourLassoModelMetrics.rootMeanSquaredError
```

Here is the output:

```
7.030519540791776
```

We have found the root-mean-square error.

Index

© Raju Kumar Mishra 2018
R. K. Mishra, *PySpark Recipes*, https://doi.org/10.1007/978-1-4842-3141-8

Get the eBook for only $5!

Why limit yourself?

With most of our titles available in both PDF and ePUB format, you can access your content wherever and however you wish—on your PC, phone, tablet, or reader.

Since you've purchased this print book, we are happy to offer you the eBook for just $5.

To learn more, go to http://www.apress.com/companion or contact support@apress.com.

Apress®

All Apress eBooks are subject to copyright. All rights are reserved by the Publisher, whether the whole or part of the material is concerned, specifically the rights of translation, reprinting, reuse of illustrations, recitation, broadcasting, reproduction on microfilms or in any other physical way, and transmission or information storage and retrieval, electronic adaptation, computer software, or by similar or dissimilar methodology now known or hereafter developed. Exempted from this legal reservation are brief excerpts in connection with reviews or scholarly analysis or material supplied specifically for the purpose of being entered and executed on a computer system, for exclusive use by the purchaser of the work. Duplication of this publication or parts thereof is permitted only under the provisions of the Copyright Law of the Publisher's location, in its current version, and permission for use must always be obtained from Springer. Permissions for use may be obtained through RightsLink at the Copyright Clearance Center. Violations are liable to prosecution under the respective Copyright Law.

Printed in the United States
By Bookmasters